TECHNOLOGY SERIES   Technical memorandum No. 7

Small-scale maize milling

Prepared under the joint auspices of the International Labour Office and the United Nations Industrial Development Organisation

International Labour Office   Geneva

ILO
UNIDO
Small-scale maize milling
Geneva, International Labour Office, 1984. Technology Series, Technical Memorandum No. 7

/Technical memorandum/, /Small-scale/, /Maize/, /Milling/,
/Developing country/s. 08.06.2
ISBN 92-2-103640-5
ISSN 0252-2004
Also published in French: *Production de farine de maïs à petite échelle* (ISBN
92-2-203640-9), Geneva, 1990

ILO Cataloguing in Publication Data

A C K N O W L E D G E M E N T S

The publication of this technical memorandum has been made possible by a grant from the Swedish International Development Authority (SIDA). The International Labour Office and the United Nations Industrial Development Organisation gratefully acknowledge this generous support.

They also acknowledge permission received to reproduce in this monograph figures from other publications as follows:

Figures	Publication
I.1	Intermediate Technology Development
III.1-4, III.6, III.10-11, III.13	Group : Tools for agriculture, a buyer's guide to low-cost agricultural implements
IV.1-2, IV.4, IV.12-13 IV.15-16	compiled by John Boyd (London, Intermediate Technology Publications, 1971)
III.5, III.12 IV.7-10	Commonwealth Secretariat : Guide to technology transfer in East, Central and Southern Africa (London, 1981)
III.9 IV.3, IV.11, IV.14	Food and Agriculture Organization of the United Nations: Processing and storage of foodgrains by rural families (Rome, 1979)

P R E F A C E

The processing of food grain plays an economically important role in developing countries for the following two reasons. First, processed grain is one of the most important elements in the diet of low-income groups, especially in urban areas where these groups are not equipped to carry out the basic processing of agricultural and animal products. Second, the use of appropriate technologies for the processing of grain may contribute to the achievement of important socio-economic objectives such as employment generation and the saving of scarce foreign exchange. Consequently the expansion of food grain processing should be carefully planned with a view to maximising the benefits which may be derived from it.

A number of developing countries have developed and promoted food grain processing techniques with a view to maintaining an adequate balance between small units, using labour-intensive or intermediate techniques (e.g. small plate, stone or hammer mills) and large units using imported, capital-intensive techniques (e.g. roller mills). Other countries have not been able to maintain such a balance and have favoured the establishment of large-scale milling plants to the detriment of small mills.

One factor which may explain this reliance on imported technology is the lack of technical and socio-economic information on alternative milling technologies. This lack of information also applies to other food processing sectors such as oil extraction, fruit and vegetable processing, fish processing and so on. Public planners and project evaluators from industrial development agencies have therefore tended to favour large-scale technologies for which information was readily available from equipment suppliers and engineering firms since they did not have information on small-scale technologies which were not widely used. Consequently, the International Labour Office has started a new series of technical memoranda on food processing technologies for wide dissemination among small-scale producers as well as public planners and industrial development agencies which have an important role to play in the promotion of these technologies. Eight such technical memoranda have already been published or are under preparation.[1] Some of these memoranda are jointly published with the United Nations Industrial Development Organisation (UNIDO) and the Food and Agriculture Organisation of the United Nations (FAO).

This technical memorandum on maize miling represents a classical example of the choice of food processing technology. The issues raised relate to a wide range of factors which determine technological choice. These factors include the nutritional content of various qualities of maize meal, the location of mills, the transport of grain and product, the different shelf-lives of various types of meal, the employment effect of alternative milling techniques, the use of foreign exchange for the import of equipment, the organisation of production in different types of mills, the generation of backward and forward linkages, the marketing of maize meal, the prevailing legislation regarding the retail prices of meal and the supply of maize grain to various types of mills, and so on. These factors must be considered simultaneously when assessing the overall socio-economic impact of alternative milling technologies. Given the importance of such assessment, Chapter I is devoted exclusively to an in-depth analysis of these factors and suggests general guidelines for the formulation and implementation of various measures in favour of appropriate milling technologies. That chapter is of particular interest to public planners and industrial development agencies since they are directly concerned by the formulation and implementation of the above measures. It may be noted that although the chapter is exclusively concerned with maize milling, the issues raised also apply to any other type of food grain. Thus, the information contained in Chapter I may also be used for the formulation of similar measures concerning a large range of food grain.

---

[1] Two technical memoranda on food processing (fish processing and oil extraction from groundnuts and copra) have already been published. Five other memoranda - in addition to this one - are at various stages of preparation.

The remaining chapters of the memorandum - Chapters II to VI - are of a mostly technical nature, and will be of particular interest to established or would-be millers. They provide detailed technical information on various aspects of maize milling, including grain preparation, shelling and milling.

The memorandum covers a narrow range of maize products, namely whole meal - produced by small-scale custom and merchant mills - and bolted and super-sifted meals produced by roller mills. Detailed technical information is mostly provided on milling technologies used in the former mills, since the detailed description of large-scale roller mills is outside the scope of this memorandum. Potential millers wishing to obtain information on roller mills should consult equipment suppliers listed in Appendix I of this memorandum.

The scales of production analysed in this memorandum range from one tonne to eight tonnes of maize per 8-hour day. These outputs are typical of those of custom mills and small-scale merchant mills. Roller mills with outputs exceeding 50 tonnes per 24-hour days are briefly described for the benefit of millers who may be considering investing in such mills.

This memorandum does not cover milling techniques used at the household level since its main target audience comprises, in addition to public planners and project evaluators, established or potential entrepreneurs planning to invest in a small-scale custom or merchant mill, or to improve an existing mill. Thus, manual milling equipment used by individual households or groups of households is not described.

Chapter V provides some general guidelines for an assessment of the requirements for various types of mills. Both material requirements (including the subject of floor areas) and skill requirements are covered. Chapter VI describes a methodological framework for the estimation of unit production costs with a view to enabling the reader to evaluate alternative milling technologies and to identify the one which is the most appropriate to prevailing local conditions.

Information and drawings of milling equipment of interest to small-scale producers are provided in Chapters II to IV. These drawings are not, however, detailed enough to enable this equipment to be produced by local workshops. Furthermore, most of the pieces of equipment described are patented. Local manufacturers will therefore need to enter into an agreement with equipment suppliers if they wish to manufacture or assemble shelling or milling equipment. A list of these suppliers is provided in Appendix I. That list is not exhaustive and the reader is urged to obtain information on milling equipment from as many suppliers or manufacturers as feasible. It must be stated, in this context, that the supply of names of equipment manufacturers by the ILO does not imply a special endorsement of those manufacturers by the Organisation, and that the names listed in Appendix I are a random selection.

The local manufacture of small-scale shelling and milling equipment is being undertaken by a number of developing countries. Any country should, a priori, be able to produce its own equipment economically as long as there is a sufficiently large market for it. Local manufacturers may use a number of available designs which may be obtained free of charge or for a relatively low fee. The only component which may need to be imported is the mill's or sheller's engine (electric or diesel-powered). The local manufacture of local milling equipment should be profitable from both the private and the social point of view since such manufacture will save scarce foreign exchange while generating employment (see Chapter I).

This memorandum does not describe all existing milling technologies. Rather, a choice has been made from among those which have been successfully applied by small-scale millers in a number of developing countries. Other technologies, not described in this memorandum, may also be adapted to local conditions and tried out in a few production units with a view to assessing their technical and economic efficiency. The bibliography in Appendix III lists the sources of information on those technologies.

Comments and observations on the content and usefulness of this publication can conveniently be sent to the ILO or UNIDO by replying to the questionnaire reproduced at the end of the memorandum. They will be taken into consideration in the preparation of additional technical memoranda.

This memorandum was prepared by J. Crabtree and G. Flynn (consultants currently working for the Tropical Products Institute in the United Kingdom) and M. Allal, staff member in charge of the series of technical memoranda within the Technology and Employment Branch of the ILO.

A. S. Bhalla,

Chief,

Technology and Employment Branch.

CONTENTS

CHAPTER I

ELEMENTS OF TECHNOLOGICAL CHOICE
IN MAIZE MILLING

The choice of maize milling technology represents a classical example of technological choice in food processing. It involves a large number of variables including those typically used in the evaluation of alternative technologies (e.g. wages, depreciation costs, socio-economic objectives) as well as additional factors such as product choice, transport costs and supply of raw materials. It is therefore important to describe the overall framework for technological choice in maize milling with a view to facilitating the evaluation of the technologies described in this memorandum.

This chapter should be of interest to both public planners and producers since it takes into consideration socio-economic objectives as well as factors which influence the private profitability of maize milling units.

The chapter is organised in the following manner. An analysis of the demand for various types and qualities of meal, together with a number of suggestions for influencing such demand, is followed by a section which looks into the supply of maize products in terms of available milling techniques and scales of production. A third section provides some information on the socio-economic implications of alternative technologies and suggests policy guidelines for the promotion of appropriate maize milling techniques and products.

I. DEMAND FOR MAIZE MEAL: PRODUCT CHARACTERISTICS
AND LOCATION OF CONSUMPTION

Three main types of maize meal are marketed in developing countries: whole meal; partly de-germed meal (i.e. meal from which part of the bran and germ has been removed) which is designated under various names (e.g. partly sifted meal, bolted meal, roller meal (Zambia)); and fully de-germed meal from which most of the bran and germ have been removed and which is also designated as "super-sifted meal". Quality differences exist within each type of meal, depending upon the milling technique adopted, the quality of the grain, and the addition of various vitamins.

A number of by-products may be produced through the milling process: limited amounts of poultry feed (coarse bran), cattle feed, and maize oil through the further processing of the germ that has been removed.

I.1 Nutritional elements

Maize is an important element in the diet of the population of developing countries, especially in Africa and Latin America. In some cases (e.g. Malawi), maize may account for 80 to 90 per cent of the total calorie intake of the rural population. Yet maize is deficient in a number of essential nutrients, and excessive reliance on maize diets could result in severe diseases such as pellagra. Consequently, a number of countries have implemented various measures for the enrichment of maize meal for human consumption (e.g. addition of various vitamins, soya bean meal, groundnut flour).

Given the deficiency of maize in a number of essential nutrients, it is paradoxical that a number of developing countries have allowed or even favoured the adoption of milling technologies which further reduce the amount of these nutrients per unit of output as well as the total amount of maize meal for human consumption per tonne of processed maize. Reasons for the increasing adoption of these technologies will be provided later, and suggestions will be made for the promotion of milling techniques consonant with desirable socio-economic objectives. It will first be useful to indicate the effect of alternative milling technologies on the nutritional value of the meal produced.

Table I.1 provides estimates of the main nutrients which constitute the three types of maize meal consumed in developing countries: whole meal, bolted meal (partly de-germed) and de-germed (super-sifted) meal. While the number of calories per 100 grams of flour are approximately equal for the three types of meal, the proportions of important nutrients per unit weight of flour are generally much larger for whole meal than for bolted and super-sifted meal. This is particularly true for calcium, iron, niacin, riboflavin and the fat content of the meal. The production of sifted meal by roller mills also removes two important types of proteins (globulins and glutelins), leaving zein which is a poorer source of protein.[1] Furthermore, enriched sifted meal is in general nutritionally less adequate than whole meal.

Table I.1

Nutrient composition of different types of maize meal

Product/ Nutrient	Whole meal	Bolted meal (partly de-germed)	De-germed meal (super-sifted)
Moisture (percentage)	12-13	12-13	12-13
Calories per 100 gr	353-356	360	363
Protein (percentage)	9.3-9.5	9.3	7.9-8.4
Fat (percentage)	3.8-4.5	Variable (>2)	1.2-2.0
Carbohydrates (percentage)	73.4	Variable (>74)	78.4
Fibre (percentage)	1.9-3.0	0.7-1.0	0.6-0.7
Ash (percentage)	1.3	n.a.	0.5
Calcium (mg per 100 gr)	7-17	6	5-6
Iron (mg per 100 gr)	2.3-4.2	1.8	1.1-1.2
Thiamine (mg per 100 gr)	0.3-0.45	0.35	0.14-1.18
Niacin (mg per 100 gr)	1.8-2.0	1.3	0.6-1.0
Riboflavin (mg per 100 gr)	0.11	0.09	0.08

Note:Variations in data according to sources may be explained by variations in the raw materials analysed and/or variations in the quality of meal.

Sources: Schlage (1968); FAO (1968); FAO (1953); FAO (1954); Uhlig and Bhat (1979).

Table I.2 provides estimates of the minimum daily consumption of whole meal and sifted meal (de-germed) which will be required for an adult's

---

[1] Source : see Stewart (1977)

daily needs of four essential nutrients: iron, thiamine, riboflavin and
nicotinic acid. The table shows that, should adults rely exclusively on
maize, they will need to consume two to seven times more de-germed meal than
whole meal in order to satisfy their daily needs of these four nutrients.
Obviously, few people will rely exclusively on maize for their diet, and the
estimates provided in Table I.2 are of a purely theoretical nature. However,
as to be shown later, the deficiency of de-germed meal in a number of
essential nutrients could have serious repercussions on the diet of low-income
groups in developing countries.

Table I.2

Minimum daily consumption of whole meal and super-sifted meal
(Amounts of meal required in grams to provide an adult's daily
needs of one mineral and three vitamins)

Substances Maize meal	Iron	Thiamine	Riboflavin	Nicotinic acid
Wholemeal	233	357	1,076	1,066
Sifted meal	424	2,500	4,670	2,670

Source : FAO (1968); Harper (1974).

Milling techniques also affect the availability of flour for human
consumption. The extraction rates of bolted or sifted meal per tonne of
processed maize are much lower than those of whole meal. Table I.3 shows that
extraction rates for whole meal vary from 97 per cent to 99 per cent while
those for bolted and sifted meal vary respectively from 80 per cent to 96 per
cent and from 60 per cent to 75 per cent. On the other hand, the proportion
of by-products may be as high as 40 per cent in the case of sifted meal
whereas generally it does not exceed 3 per cent for whole meal. These
by-products are used either as animal feed or for oil extraction. The
production of bolted or sifted meal may thus have important repercussions on
consumption by low-income groups for the following reasons.

First, from the purely quantitative point of view, and given the above extraction rates, approximately 20 to 40 per cent of cultivated maize will not be available for direct human consumption if, for example, it is used to produce sifted meal instead of whole meal (the corresponding percentages for bolted meal will be 3 to 17 per cent). The lower extraction rates associated with the production of sifted and bolted meal should be of little consequence for countries producing a net surplus of maize (e.g. for export or further processing into oil or animal feed). On the other hand, countries which do not grow enough maize to fully satisfy the needs of their population may face larger shortages of maize meal if sifted rather than whole meal is produced. If these countries do not make up for these shortages by importing maize or other kinds of grain, in the absence of price controls the low-income groups will either have to pay higher prices for the available supply of maize or cut down consumption. In either case as shown in table I.3, they will suffer from the lower supply of maize flour due to the low extraction rates of sifted maize mills (i.e. roller mills).

Table I.3

Extraction rates per 100 kg of maize
(percentages)

Product \ Use	Flour for human consumption	By-product used for animal feed or oil extraction
Whole meal	97-99	1-3
Bolted meal	80-96	4-20
Sifted meal	60-75	25-40

Source : Stewart (1977) and Uhlig and Bhat (1979).

Secondly, the use of by-products (bran and germ) as animal feed (e.g. for cattle or poultry) may not fully compensate for the low extraction rates associated with the production of sifted meal: if these by-products are used to increase the production of meat, their nutritional value will not generally exceed 10 per cent of that of bran and germ (i.e. the conversion of bran and germ into meat results in a 90 per cent loss of the nutritional value of these two by-products). Furthermore, it is very unlikely that low-income groups will be able to afford to improve their daily diet by buying the meat produced.

Thirdly, it may be noted that by-products are generally locally marketed or exported as animal feed and that few developing countries have maize oil extraction plants. In any event, however, the extraction of oil from the by-products may not justify the production of sifted meal. Two important factors should be considered before deciding to invest in a maize oil extraction plant. These are:

- whether a sufficiently large and regular supply of by-products will be available to ensure a continuous high capacity utilisation of the plant, and therefore competitive (non-subsidised) retail prices of oil; and
- whether oil should not be obtained from some other raw material given the importance of maize for direct human consumption.

The previous analysis of the nutritional value of the three types of maize meal indicates that in terms both of quality and quantity, the production of whole meal is to be preferred to that of sifted or bolted meal if the satisfaction of the basic needs of low-income groups constitutes a major development objective. Other factors may, however, vitiate the above analysis, as indicated below.

I.2 Shelf-lives

An important consideration in the marketing of maize meal is its shelf-life. A reason often advanced for the production of sifted meal (and to a lesser extent bolted meal) is that its shelf-life is considerably longer than that of whole meal. Thus, in cases where the marketing chain requires long shelf-lives (e.g. when the meal must be transported over long distances or when wholesalers and retailers must keep stocks for an extended period), the only practical possibility is to produce sifted meal because whole meal tends to become quickly rancid as a result of its high fat content (approximately 3 to 4 per cent fat as against 1 to 2 per cent for sifted meal). This reason for justifying the production of sifted meal has not, however, met with the unanimous agreement of practising millers and retailers of maize meal in developing countries. Various aspects of this matter are briefly outlined below.

Estimated shelf-lives

The shelf-life of maize meal depends on the following: the fat content of the meal, the humidity of the maize grain, the presence of various contaminants in the maize and the meal storage conditions (e.g. choice of packaging material, air temperature and humidity level within the storage area). The effect of the above factors on the shelf-life of maize meal has not yet been investigated in a systematic manner. Differences in shelf-life between sifted meal and whole meal may not, therefore, be ascribed entirely to differences in fat content. The differences in shelf-life could well be explained to a certain extent by the conditions under which maize is processed in roller mills producing sifted meal and in hammer or stone mills producing whole meal (e.g. humidity level of the maize, presence of contaminants). The widely divergent estimates of shelf-lives found in publications on the subject may thus be explained by big differences in the above conditions from one country to another or among processing units within the same country. Estimates of these shelf-lives are provided below:

- 4-6 weeks for whole meal if stored at high temperatures and humidity levels, as compared to a maximum of two years for sifted meal if stored under carefully controlled conditions (Uhlig and Bhat, 1979);

- 2-3 days for whole meal versus a considerably longer shelf-life for sifted meal (Stewart, 1977);

- 4-8 weeks for whole meal versus 6 months for sifted meal, but humid or hot climates do not allow long storage of any type of flour (JASPA, 1981);

The JASPA (1981) study also includes estimates provided by individual millers in Kenya and Zambia:

- sifted flour may last up to 3 months if the moisture content
  can be kept at a low level, while a 5 per cent fat wholemeal
  flour will last for only three weeks (a miller in Kenya);
- storage characteristics are almost identical for all types
  of flour, with shelf-lives reduced to 2 weeks during the
  rainy season (manager of a large mill in Zambia);
- there are few differences between the shelf-lives of sifted
  meal and whole meal (small-scale miller in Kenya).

The above views on the shelf-lives of sifted flour and whole meal indicate
that this matter needs to be further investigated by food technologists in
developing countries with a view to determining the extent to which
differences in shelf-lives do exist, the reasons for such differences, and
whether it will be possible to promote appropriate milling conditions (e.g.
use of sufficiently dried maize, adequate storage conditions) with a view to
narrowing down the difference between the shelf-lives of whole meal and sifted
meal.

Need for extended shelf-life of maize meal

Although there is a consensus regarding the longer shelf-life of sifted
flour, no such consensus seems to exist regarding the importance and
significance of the difference in shelf-lives between the two types of meal.
It is argued that consumers in rural areas - especially maize-growing areas -
have their maize milled to order as the need arises, and do not generally
stock whole meal for more than a week. It is further argued that long
shelf-lives will be required mostly in large urban areas where flour must be
stored over an extended period because large stocks are needed in order to
ensure steady and regular supplies of flour to retailers. Sifted flour may
also be required wherever there is a need to transport it over long distances
to parts of the country where maize is not grown.

This line of reasoning, which assumes different shelf-lives for sifted
flour and whole meal, may need to be modified for the following reasons.
First, whole meal can be produced by small merchant mills in urban areas if
the grain can be supplied to them on a regular basis by, for example, a
national maize marketing board. Thus, whole meal can be produced and retailed

in such a way as to reduce the storage period to a minimum and therefore avoid rancidity. In this case, it will be marketed as a perishable food product but with a shelf-life longer than that of meat or milk, for example. Secondly, custom or merchant mills can also be established in parts of the country where maize is not grown as long as the transport of shelled maize can be ensured on a regular basis. However, it must be emphasised that long storage of shelled maize for further processing by small custom or merchant mills will require that the maize be properly dried (i.e. moisture level not to exceed 12 to 13 per cent) and stored, and that adequate transport facilities be available.

I.3 Consumers' preference, retail prices and marketing channels

Demand for whole meal, bolted meal and sifted flour is function of three main factors: consumers' preference, retail prices of the different kinds of maize meal and marketing channels. These factors are briefly analysed in the light of available evidence from developing countries, especially in the African region.

Consumers' preference

It is argued that if they had a choice, consumers in developing countries would prefer to buy sifted (i.e. de-germed) flour rather than whole meal for a number of reasons, including the better appearance of sifted flour (it is whiter and finer than whole meal) and its easier utilisation in cooking (whole meal needs more energy than sifted flour, and thus takes more time to cook). On the other hand, consumers do not seem to be concerned - probably for lack of information - by the lower nutritional value of sifted flour. The above arguments do not, however, apply in all cases. In some countries (e.g. Somalia - see JASPA report, 1981), the urban population prefers whole meal whenever it is available. Large institutions such as hospitals also tend to prefer the more nutritious whole meal. Available evidence on this subject is generally limited, and it will be useful for developing countries to undertake individual surveys of consumers' preference for various types of maize meal.

For some urban areas (e.g. in Kenya) where preference for sifted flour is well established, a number of both objective and subjective reasons may explain that preference. The objective reasons include those already mentioned (i.e. attractive appearance of sifted flour, its better cooking characteristics) as well as the possibility of buying limited amounts of packaged flour at food stores close by. The apparently longer shelf-life of sifted flour does not seem to play a role in the urban consumers' choice, probably because the flour is used up in a relatively short time.

Among the subjective reasons adduced for the preference for sifted flour by urban consumers, advertising is by far the most important. Available evidence from a number of African countries (e.g. Kenya) shows that large milling firms (some of them of foreign origin) earmark large sums for major advertising campaigns in urban areas. For example, collected information from Kenya shows that in some cases marketing costs amount to as much as a third of production costs (Stewart, 1977). It is even suggested that the market for sifted flour might have been artificially created by the large roller mills (see Stewart, 1977).

Retail prices

Available evidence shows that the retail price of sifted meal is generally higher than that of whole meal, price differences varying from country to country as well as within the same country. This does not mean that the actual milling costs of sifted meal are necessarily higher than those of whole meal. Differences in retail prices may also be the result of the following factors:

1. Low extraction rates of roller mills which produce the sifted flour (i.e. less flour is produced per tonne of shelled maize than in the case of the production of whole meal). These low extraction rates, and therefore higher input of raw materials is not offset by the sale of by-products (germ and bran) since the unit prices of the by-products are generally much lower than that of whole meal;

2. The high packaging cost of sifted flour (e.g. in 1 kg or 2 kg paper bags) whereas little if any packaging is used in the case of whole meal. Custom mills (generally located in rural areas) do not package their products (the customers bring their own containers) while small merchant mills use cheap packaging;

3. Roller mills market their product through traders who must add their operating costs and profit margin to the ex-mill price.

4. For sifted meal, there are high advertising costs whereas there is no costly advertising for whole meal;

5. Transport costs associated with the production and marketing of sifted meal are generally higher than those associated with that of whole meal. Maize and sifted meal must generally be transported over long distances, whereas transport costs are cut to a minimum by the proximity of custom mills (which produce whole meal) to maize-growing areas and to the consumers.

The relatively high retail prices of sifted meal do, generally, limit consumption to the middle-income and high-income groups in urban areas. In a few exceptional cases, government subsidies and price controls have maintained retail prices of sifted meal low enough to allow consumption by low income groups.

In general, an increase in the retail price of sifted flour should not lead to an increased demand for whole meal on the part of middle-income and high-income groups. It may rather lead to an increase in demand for polished rice or other flours (e.g. wheat flour) of equivalent quality should these be locally available. On the other hand, an increase in the retail price of wholemeal maize should generally increase demand for other meal of equivalent quality (e.g. sorgho meal, millet) by low-income groups . Although no firm evidence exists on the above shifts in demand, their possibility should be taken into consideration whenever government action may lead to an increase in the retail price of sifted maize flour or to a decrease in its availability for whatever reason.

Geographical distribution of demand
for sifted flour and whole meal

As stated earlier, rural areas consume almost exclusively whole meal, especially in maize-growing areas. Furthermore, whole meal is either produced by households (e.g. by the use of mortar and pestle or of querns), or by custom mills for payment in cash or kind. Recourse to custom mills is not yet widespread. However, an increasing number of rural women are willing to give up the long and tedious milling of maize at home in order to be able to devote themselves to other more profitable activities.

It is doubtful whether demand for sifted flour will expand substantially
in maize-growing areas for the foreseeable future. The retail price of sifted
meal will generally not be competitive with that of whole meal whether the
latter is produced at home or at a custom mill. On the other hand, sifted
meal is in some cases marketed in rural areas which do not grow maize,
especially if no other cereals are available. In some cases, the sifted meal
is sold at subsidised prices because it could not be otherwise afforded by the
poor in rural areas.

The situation in urban areas is different: both whole meal and sifted meal
are consumed by the urban population. Whole meal is often produced by small
merchant mills (e.g. hammer mills) which keep stocks of both the raw materials
and the meal. These mills either retail their output directly or sell it to
retailers. Some whole meal is also produced in neighbouring rural areas and
sold in the urban markets. In general, the whole meal marketed in urban areas
is consumed by low-income groups.

Sifted meal is in most cases marketed in urban centres. It is produced by
large roller mills located in urban areas, or, if feasible, close to the
maize-growing areas. Sifted meal is mostly consumed by the richer sections of
the urban population and is marketed through traders. Consumption of sifted
meal by the poor is rather limited since its higher price puts it in the
luxury goods category.

II. SUPPLY OF MAIZE MEAL: MILLING TECHNIQUES AND SCALES OF PRODUCTION

The five distinct milling techniques used in developing countries are
listed below:

(1) The mortar and pestle technique used in the household. Output per
hour varies from one person to another, but does not generally exceed 5 kg per
hour.

(2) Hand-operated grinders used by individual households or groups of
households. Available equipment allows production rates ranging from 7 kg to
30 kg per hour. Figure I.1 shows one type of hand-operated mill produced in
Kenya.

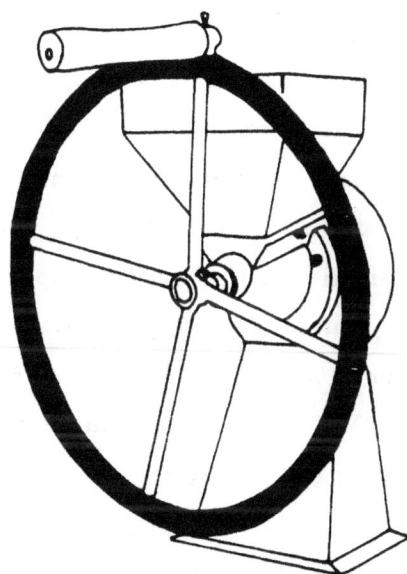

This small, hand-operated plate mill is of
all-welded steel construction and has been
designed in Africa **for** the grinding of maize
and other food grain . The front cover can be
removed for cleaning and inspection by un-
screwing three wing nuts.

<u>Source</u> :   ITDG (1976)

<u>Figure I.1</u>

Duna hand-operated grinding mill

(3) Water-powered stone mills, mostly used in the rural areas of a number of African countries. Their production rates are relatively low, varying from 20 to 30 kg per hour, depending on the force of the water flow. These mills usually operate as custom mills.

(4) Engine-powered hammer mills and stone mills, equipped with diesel or electric engines, and used by custom mills and merchant mills for the production of whole meal. The rated outputs for these mills range from about 100 kg/hr to 1,100 kg/hr. These outputs cannot, however, be maintained over long periods, capacity utilisation ranging usually between 200 to 3,000 kg per 8-hour day.

(5) Roller mills of various sizes producing partly or fully de-germed meal, and generally located in urban centres. Their output varies on average from 1 tonne to 12 tonnes per hour, depending on the mill's size, the number of shifts worked and the fraction of time used for maintenance and repairs. In developing countries, roller mills generally operate at lower capacity levels than in industrialised countries where the economic size of roller mills ranges between 250 tonnes and 300 tonnes per three-shift day. By comparison, an output of 120 - 150 tonnes per three-shift day is relatively large for roller mills in developing countries.

An increasing number of developing countries (e.g. India, Kenya, Tanzania) produce hand-operated grinders, water-powered stone mills and small hammer mills. Roller mills, on the other hand, are mostly produced in industrialised countries, India being one of the few developing countries which produce small roller mills.

Roller mill equipment does not differ much from one equipment manufacturer to another, the main differences relating to ancillary operations such as the feeding of maize and packaging. From economic comparisons undertaken in Kenya by Uhlig and Bhat (1979), it would seem that, given the relatively low wage rates prevailing in developing countries, manual or semi-automated feeding and packaging are more appropriate than fully automated ancillary operations.

Regarding hammer mills, a number of studies (Uhlig and Bhat (1979), Stewart (1977), JASPA (1981)), have shown that equipment used in developing countries can be improved with a view to increasing productivity and decreasing maintenance costs. For example, hammer mills can be fitted with magnets in order to remove pieces of metal in the grain which might break the sifting gauge. The hammers used in some of the mills can also be re-designed in order to improve the meal quality and the productivity of the equipment. For example, some hammer mills produced in a developing country make use of fixed hammers (within the grinding chamber) which are not as efficient as non-fixed hammers (see R. Kaplinsky in Baron, 1980).

The capacity utilisation of hammer mills and roller mills varies from country to country. In general, hammer mills are used at high capacity levels because they are fairly versatile and can be used, with few adjustments, for the grinding of a number of cereals in addition to maize. Thus, they can still operate outside the maize-growing season. Roller mills, on the other hand, are used exclusively for the processing of a single grain (e.g. maize) and must depend on large stocks of raw materials for continuous running. Thus the lack of sufficient stocks is often responsible for the low capacity utilisation of roller mills and has at times necessitated government action to ensure sufficient supplies. For example, it is reported in JASPA (1981) that the government of an African country has taken steps to ensure the priority supply of maize to a government roller mill to the detriment of small hammer mills.

While the capacity utilisation of rural hammer mills is relatively high, it is reported (see JASPA, 1981), that some small urban mills may be suffering from the competition of roller mills in areas where consumers' preference has shifted in favour of sifted meal. However, the effect of such a shift on the profitability of small urban mills is not precisely known.

III. CHOICE OF MAIZE MILLING TECHNOLOGIES AND DEVELOPMENT OBJECTIVES

The previous two sections provided an overall assessment of demand for and supply of different types of maize meal in developing countries. It was also suggested that existing demand and supply conditions could be improved for the benefit of both consumers and producers. In addition, like any other sector of production, maize milling can be organised in such a way as to contribute to the government's socio-economic objectives. The present section therefore puts forward some tentative guidelines for the identification and adoption of milling technologies consonant with adopted development objectives. Although these guidelines apply specifically to maize milling, they can also be applied, with suitable adjustments, to other food grains.

Development objectives may vary from country to country according to prevailing local socio-economic conditions. Consequently, the choice of maize milling techniques may also vary according to these conditions. However, existing development plans and available evidence tend to suggest that the following objectives - as they relate to the production and consumption of maize meal - are common to the majority of developing countries.

Self-sufficiency in food.

Food is one of the most important basic needs, and a main objective of developing countries is to expand the local production of food products, especially in favour of low-income groups. Since the purchasing power of these groups is relatively limited, priority should be given to the production of low-priced food products with a high nutritional value while making due allowance for consumers' preferences. If some sections of the population (e.g. the better-off and the urban population in general) require other types of foods (e.g. sifted flour instead of whole meal), the government may satisfy such demand either through limited local production or by imports. In the case of maize, the production and marketing of whole meal seems to be more appropriate than that of sifted meal given the high nutritional value and relatively low prices of whole meal. Demand for sifted meal may be satisfied through limited production by local roller mills as long as such production does not jeopardise that of whole meal.

Technological self-reliance.

Another development objective of a large number of developing countries is to promote technological self-reliance with a view to reducing the costly import of know-how and equipment, and promoting the production of consumer and capital goods consonant with the local needs, customs and culture. In the case of maize milling, the promotion of hammer mills or hand-operated mills constitutes, for the majority of developing countries, the right step towards the achievement of technological self-reliance. This is not the case for roller mills which must be imported by most developing countries, India being one of the few countries which have mastered the roller mill technology.

Employment generation

Employment generation is, by far, one of the most important development objectives. A number of developing countries have focussed their efforts and resources towards the achievement of this objective with special emphasis on the promotion of rural employment.

Regarding maize milling, the employment generation objective may take into consideration the following:

- the total direct employment generated per unit of output;

- the investment cost per unit of employment generated; this criterion is of paramount importance for countries suffering from severe shortages of local capital and foreign exchange;

- skill requirements per unit of output; this factor is also quite important, since the shortage of skilled labour requires the implementation of long and costly training programmes;

- indirect employment effects, such as those generated by the transport of maize grain, of whole meal or sifted flour, the production and maintenance of milling equipment and the production of packaging material;

- foreign exchange savings; and

- rural industrialisation.

The above factors are analysed below in relation to the existing maize milling technologies.

Total direct employment per unit of output

The available evidence does not yield reliable estimates on the employment effect of alternative technologies. Table I.4 provides estimates of output per man-hour from Stewart (1977) and JASPA (1981).

Table I.4

Labour intensity of alternative milling techniques

(in tonnes per work-hour)

Mill \ Author	Stewart 1977	JASPA 1981
Water-powered stone mill	0.018	-
Hammer mill	0.198 - 0.225	0.041 - 0.062
Roller mill	0.153	0.040 - 0.093

Table I.4 shows that the labour intensity of hammer mills is either lower or higher than that of roller mills, depending on the source of information. Differences in scales of production, capacity utilisation, or levels of automation (e.g. fully or partially automated roller mills) may explain the different estimates obtained by Stewart and JASPA.

The fact that hammer mills may not be much more labour-intensive than roller mills is not altogether surprising for the following reasons. First, the roller mill technology includes a number of sub-processes which are not generally part of the hammer mill technology (e.g. grain cleaning, de-germing, packaging). Secondly, medium and large-scale roller mills require a much larger managerial and maintenance staff than do hammer mills, especially when the latter are custom mills. It may thus be concluded from available evidence, that the direct employment criterion does not favour any particular milling technology, the only exception being the water-operated stone mills which are particularly labour-intensive (0.018 tonnes/man-hour). However, as will be noted later, the indirect employment criterion does favour the establishment of hammer mills.

Investment cost per unit of employment

As is to be expected, this criterion favours hammer mills by a wide margin. Table I.5 provides estimates of investment costs per worker for hammer mills and roller mills.

Table I.5

Investment costs per worker

| Author | Stewart 1977 [1] | JASPA 1981 [2] |
Mill	(Eastern Africa shillings)	(Tanzania shillings)
Hammer mills	8,350 - 12,830[a]	30,800 - 38,350[b]
Roller mills	41,180[a]	131,740 - 214,425[c]

[a] One-shift operation.
[b] Two-shift operation.
[c] Three-shift operation.

Table I.5 shows that, depending on the source of information, investment costs per worker for roller mills are three to seven times higher than those for hammer mills. These estimates do not take into consideration working capital which is considerably higher for roller mills than for hammer mills.

Thus, it may be concluded that the promotion of hammer mills should benefit developing countries that are short of investible funds but wish to expand employment.

Skill requirements

Skill requirements for hammer mills are substantially lower than those for roller mills. Two to three weeks of on-the-job training are usually sufficient for the operation of hammer mills, skilled labour being needed mostly for maintenance. On the other hand, 30 to 50 per cent of the workers of roller mills are in the skilled labour category (see, for example JASPA, 1981). The establishment of hammer mills should therefore require much less training of labour and will not require the recruitment of costly foreign manpower.

Transport of raw materials and flour

Whenever production takes place in large roller mills, the maize grain and meal produced must generally be transported over long distances, for the following reasons. First, roller mills are often located in urban centres where the necessary facilities (e.g. energy) are fairly well developed and where there is a sufficiently large pool of skilled labour. The raw materials must therefore be transported from generally distant maize-growing areas to the roller mills. Secondly, as the flour produced is marketed in the main towns of the country, it also must be transported over long distances. Whole meal, on the other hand, is often produced close to the maize-growing area and thus little transport is required. For example, maize processed by custom mills is often transported over short distances by foot or by animal-drawn carts. Long-distance transport of maize to hammer mills occurs whenever they are located in areas which do not grow maize or in urban centres where they are operated as merchant mills.

The foregoing assessment of transport needs shows that in general, employment generated by such transport should be substantially larger if maize were processed in large-scale roller mills than if it were processed in small-scale hammer mills. However, because of the high cost of petroleum and transport equipment, and the need for most developing countries to import them at the expense of other essential goods, it is not always in the national interest to rely on the transport of maize grain and flour as a means of generating employment.

Production of milling equipment

The large majority of developing countries do not produce roller mills and, for small to medium-sized countries the production of such mills may never be feasible because demand will not be sufficient to justify the establishment of a manufacturing unit. On the other hand an increasing number of developing countries are now producing various types of hammer mills, and there seems to be no technical or economic reasons to prevent most countries from producing such mills. The manufacture of hammer mills should generate substantial indirect employment through the production of the various mechanical components, whereas the manufacture of roller mills will have no such advantage. However, it may be noted in passing that hammer mills run on diesel or electric engines, which developing countries generally do not produce themselves and will therefore have to import.

Estimates of the employment generated by the production of hammer or stone mills are not generally available. A JASPA (1982) study indicates that in one year ten workers can produce up to 100 small mills with imported engines and that the repair and maintenance of 30 small hammer mills need one full-time worker. Considering that, on average, one small hammer or stone mill is needed for every 1,000 inhabitants, and that 10 per cent of the mills must be replaced every year, the employment generated by the production, maintenance and repair of mills could be significant.

Other backward and forward linkages may also generate indirect employment (e.g. production of packaging materials, marketing). These are, however, of minor importance and should not significantly affect the choice of milling technology.

Foreign exchange savings.

The use of hammer or stone mills in place of roller mills should yield substantial foreign exchange savings and should therefore be of particular interest to countries suffering from balance of payments problems. Table VI.I provides a price list for a large number of stone, plate, hammer and roller mills. It can be seen that while the f.o.b prices of stone, plate and hammer mills vary between £200 and £10,000 (1980 prices) depending on the mill's output, those of roller mills vary between £250,000 and £700,000. Foreign exchange savings may be illustrated by the following example based on the price list in Chapter VI. Let us assume that a country may choose between a roller mill with an output of 120 tonnes a day on three-shift working or eight hammer mills with an individual output of 15 tonnes per day on two-shift working. The f.o.b. price of the roller mill from the United Kingdom will be £400,000 while the cost of the eight hammer mills from Brazil will be £6,000 f.o.b. If the hammer mills were to be produced locally with imported engines and steel, foreign exchange costs might not exceed £3,000.

Other similar examples from the price list will indicate that substantial foreign exchange savings may be derived from the adoption of stone or hammer mills in place of roller mills.

Rural industrialisation

A large number of developing countries have launched programmes in favour of rural industrialisation with a view to improving the employment and incomes of the rural population and slowing down their migration to urban areas. Food processing, and in particular the milling of grain, constitutes a basic rural industry. It is therefore important to maintain this type of industry in rural areas, and to avoid measures which will put processors at a disadvantage in relation to processors established in urban areas. For example, subsidisation of large roller mills or government measures which may restrict the supply of maize to small rural mills (e.g. in cases where the limited supply of maize is assigned, in priority, to roller mills by government decree) could, in the long run, force the closing down of hammer mills in rural areas. In some countries, such as Tanzania and Kenya, some small mills have already closed down or are operating at low capacity partly as a result of the expansion of roller mills.

IV. CRITERIA AND METHODS OF GOVERNMENT ACTION

The previous section suggests that, from a socio-economic point of view, the production of whole meal by small merchant or custom mills might be moreappropriate than that of sifted or bolted meal by large plants. However, as pointed out earlier, the choice of milling techniques may vary from one country to another depending on such factors as the level of development, consumers' tastes and the distribution of the population between rural and urban areas. The promotion of any particular milling technique should therefore follow a careful evaluation of the supply and demand of different kinds of maize meal, taking into consideration the country's development objectives. An overall study of the maize milling sector might include the following:

- a survey of consumers' demand for various types of meal, including the identification of the reasons which may explain demand for specific types of meal (e.g. availability, tastes, advertising, low prices, packaging, shelf-life);

- a survey of maize meal production by households, custom mills, merchant mills and roller mills, including the location of production units, milling techniques, scales of production, quality of output, marketing channels and wholesale and retail prices; and

- a socio-economic analysis of alternative maize milling techniques based on information obtained from the above surveys, taking into consideration the country's development objectives. The assessed techniques should include those currently used in the country as well as improved techniques which have been developed elsewhere.

Findings from the above study may then be translated into concrete action for the promotion of milling techniques consonant with the country's development objectives. Government action in this sector may include the following:

- formulation and application of measures to induce consumption of specific types of maize meal;

- dissemination of information on improved milling techniques;

- promotion of research and development directed to the improvement of the quality and shelf-life of whole meal, including the development of appropriate packaging required by some segments of the market; and

- formulation and application of measures to promote the right balance among various types and scales of mills, taking into consideration the consumption pattern which the government wishes to encourage.

The technical information contained in the following chapters should be useful for the undertaking of the suggested study, as well as for the formulation of government measures for the promotion of suitable milling techniques. It is also particularly intended for practising and would-be millers who wish to improve their current milling techniques or set up a new production unit.

CHAPTER II

GRAIN PREPARATION

I. INTRODUCTION

A number of operations must be carried out prior to the milling of maize grain, including the following:

- harvesting of the maize cobs;
- drying of cobs prior or after husking;
- husking;
- shelling; and
- storage of dried grain whenever necessary.

The above sequence of operations may be altered by, for example, shelling the cob after harvesting or husking, then drying the shelled grain for storage. The drying of maize before shelling is recommended whenever it must be stored over a prolonged period of time before milling since the leaf sheaths protect the grain against insect infestation and breakage. On the other hand, the drying of maize on the cob requires a longer drying time than that of shelled maize. Thus, the decision to dry the grain before or after shelling will depend to a large extent on its ultimate use.

The various operations preceding milling are briefly reviewed in this chapter, with the exception of shelling which is described in detail in Chapter III. Since this memorandum is mostly concerned with milling technologies used in small-scale production units (custom and merchant mills), a number of the operations listed above may not be carried out by these units. For example, custom mills usually process the dried grain brought by the customers, and do not engage in husking, shelling or drying of maize.

However, it is conceivable that some small-scale mills, particularly merchant mills located in urban areas, may carry out some of the operations. This is the reason for including this chapter on grain preparation. However, since a separate memorandum on grain storage and drying is also being prepared, these two operations will not be covered in detail in this chapter.

Small-scale mills may undertake the various operations which precede milling according to one of the following five sequences:

- Husking - drying - storage on the cob - shelling as required;

- Husking - drying - shelling - storage as shelled grain;

- Husking - shelling - drying - storage as shelled grain;

- Drying - husking - storage on the cob - shelling as required; and

- Drying - husking - shelling - storage as shelled grain.

Depending on circumstances, one or the other of the above sequences may be preferred. Weather conditions, availability of storage, and the duration of the storage period, the price and availability of fresh or dried cobs, etc. will generally dictate the choice of the most appropriate sequence.

II. HUSKING (REMOVAL OF LEAF SHEATHS)

The leaf sheaths (fresh or dry) may be removed manually or mechanically. In the latter case, the removal of leaf sheaths is generally part of the shelling operation, the powered maize shellers being equipped with an appropriate husking element. Mechanical husking is not generally appropriate for small-scale mills since the capacity utilisation of the husking and shelling equipment is generally too large for these mills.

Manual husking involves the stripping of leaf sheaths with either the bare fingers or with a glove equipped with a husking hook. The latter improves labour productivity while protecting the fingers. A typical husking hook is shown in figure II.1. The hook is first inserted into the leaf sheaths at the tassel end of the cob and then moved towards the cob base, thus stripping away

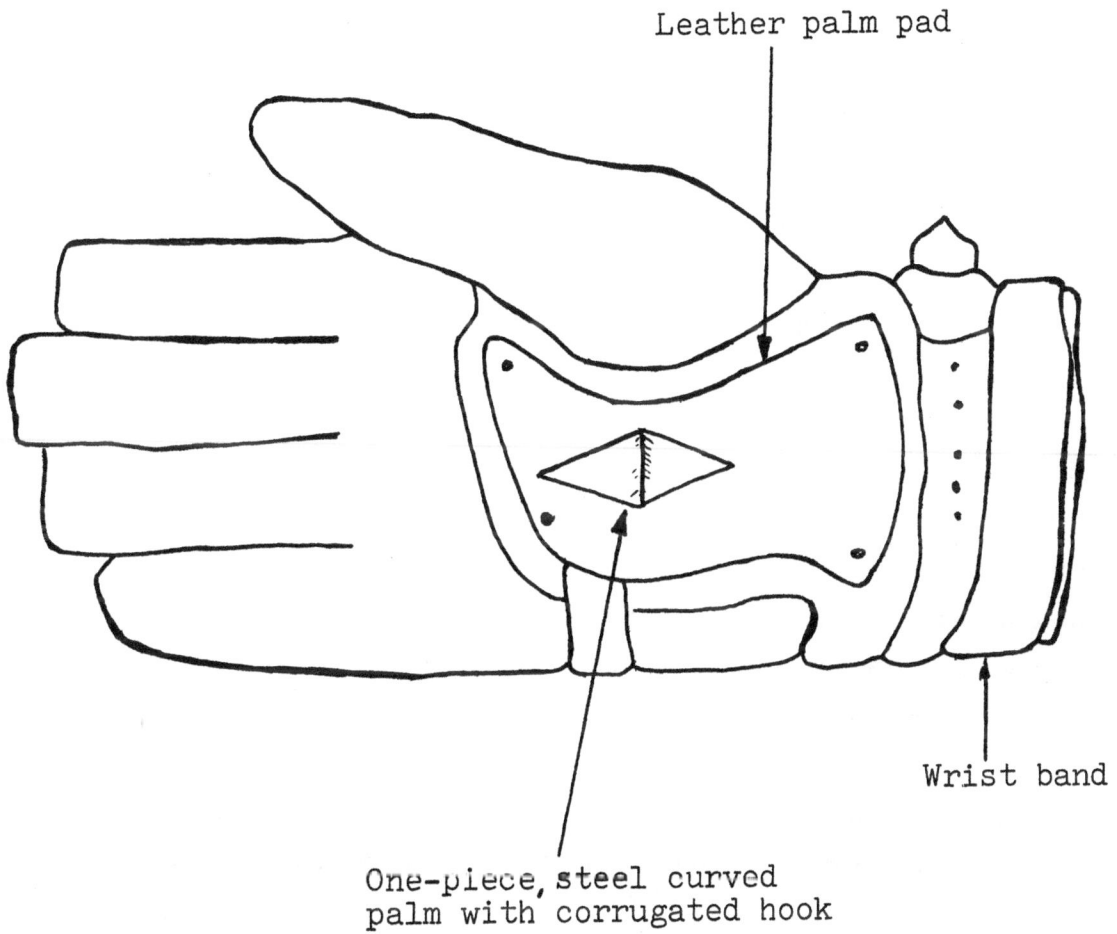

Figure II.1

A typical husking hook

the leaf sheaths from the grain. The husking hook is particularly useful when husking dried cobs since the leaft sheaths are much harder to strip away than those of fresh cobs.

III. DRYING

There is no definitive method for the drying of maize since each method depends on a number of factors, such as the level of maize production, the intended use of the grain, the capital and expertise available, fuel availability and the local weather after harvest. In this section, the principles involved in maize drying are summarised and a small selection of the available drying methods are briefly described.

Maize must be adequately dried before subsequent storage to prevent germination of the grain, the growth of micro-organisms and insect infestation. Most drying processes, either of cob maize or shelled grain, take place at or near the point of maize production.

During the drying process, moisture which evaporates from the wet grain is rapidly absorbed into the drying air until an equilibrium state is reached where no further moisture is lost. The final grain moisture content is termed the "equilibrium moisture content" for specific ambient conditions. The rate at which drying takes place depends upon the moisture content of the grain and the flow rate, temperature and humidity of the drying air.

The most important factor in the drying of grain is the air flow rate over or through the grain rather than the air temperature. However, drying may be accelerated by increasing the air temperature. The latter should be kept below some maximum value depending upon the intended end-use of the grain. For milling purposes, temperatures above $60^{o}C$ should be avoided since milling and nutritional characteristics may be adversely affected.

III.1 Moisture content determination

For safe storage, the moisture content of grain should be reduced to 13 per cent or less. It is therefore important to measure the moisture content before ending the drying cycle.

Methods of moisture content determination used in large-scale plants are complicated, expensive and inappropriate for small-scale mills. The latter should therefore rely on visual inspection and simple experiments, such as the pressing of the grain with the thumb nail or its crushing between the teeth. The drier the grain, the more resistance it offers to this pressure. An alternative, simple and effective method involves the shaking of a cupfull of shelled grain for two to three minutes in a screw-top bottle containing a tea-spoon of salt. The grain is dry enough for storage if the salt does not lump or stick to the sides of the jar (O'Kelly, 1979).

III.2 <u>Drying methods</u>

Drying of maize may be achieved by one of three basic methods or a combination of these: sun drying, solar drying and artificial drying. The drying of maize on the cob, or of shelled grain dictates the choice of drying operations in each of these methods. The main characteristics of the latter are summarised in table II.1

III.2.1 <u>Sun drying</u>

Since most areas of developing countries benefit from sufficient sunshine during the post-harvest period, dependence upon the sun for drying grain is the cheapest and most common practice amongst maize farmers.

Cob maize, either husked or unhusked, is often dried in narrow cribs with open wall construction to allow natural ventilation through the cobs. Under suitable climatic conditions, and provided that there is no excessive restriction to the air flow, the cobs dry to a safe moisture content without the development of surface mould or insect infestation. These cribs are also commonly used for the temporary storage of dried cobs.

Unhusked maize cobs are often tied by the tassels into small bundles and hung from trees or exterior house rafters to dry. Alternatively, simple racks, made from horizontal bars supported by inclined bamboo poles, may be used (see figure II.2). The cobs are hung from the horizontal bars by their tassels. In the event of a sudden shower, a similar but slightly taller rack covered with polythene sheeting or large leaves, can be placed over the first rack (see figure II.3).

Table II.1

Major characteristics of the sun, solar and artificial drying systems

	Sun	Solar	Natural convection	Artificial Forced convection	
				Ambient air	Heated air
Throughput	Very low	Moderate	Low	Moderate	High
Scale and capital cost	Very low	Moderate	Low	Moderate	High
Skill required for operation	Low	Moderate	Moderate	Moderate	High for some
Labour requirements for operation	High	Moderate	High	Low	Low
Technology and maintenance requirements	Minimal	Moderate	Low	High	High
Dependency upon climate	Total	Major	Nil	Slight	Nil
Dependency upon combustible fuel	Nil	Nil	Total	Nil	Total
Dependency upon external power (electricity)	Nil	Total	Nil	Total	Total
Control of drying	Nil	Moderate	Low	High	High
Susceptibility to grain damage	Moderate	Low	High	Low	Low
Protection against insect or microbial infestation	Low	Moderate	Moderate	Moderate	High

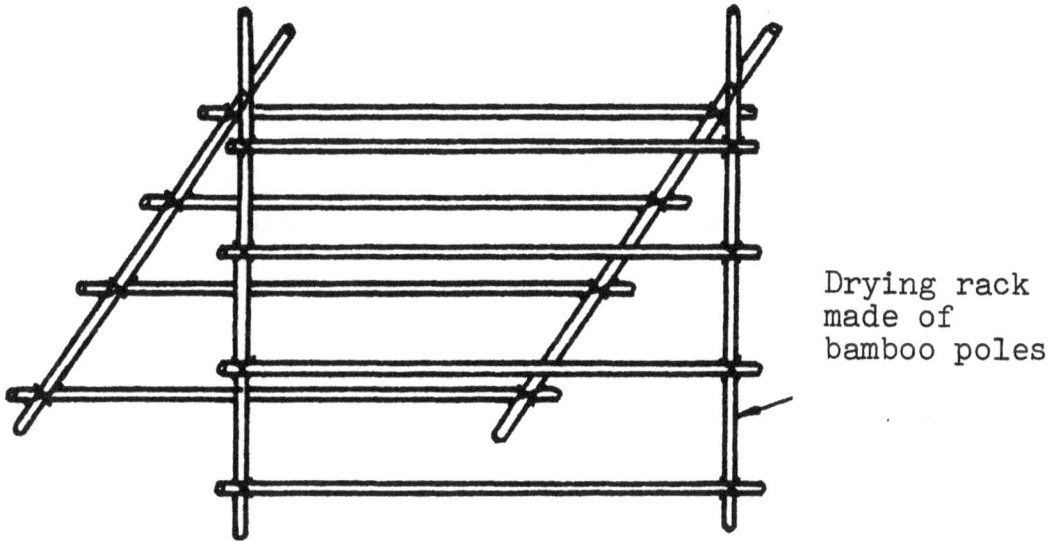

Drying rack
made of
bamboo poles

Figure II.2

Simple rack for the sun drying of
unhusked maize cobs

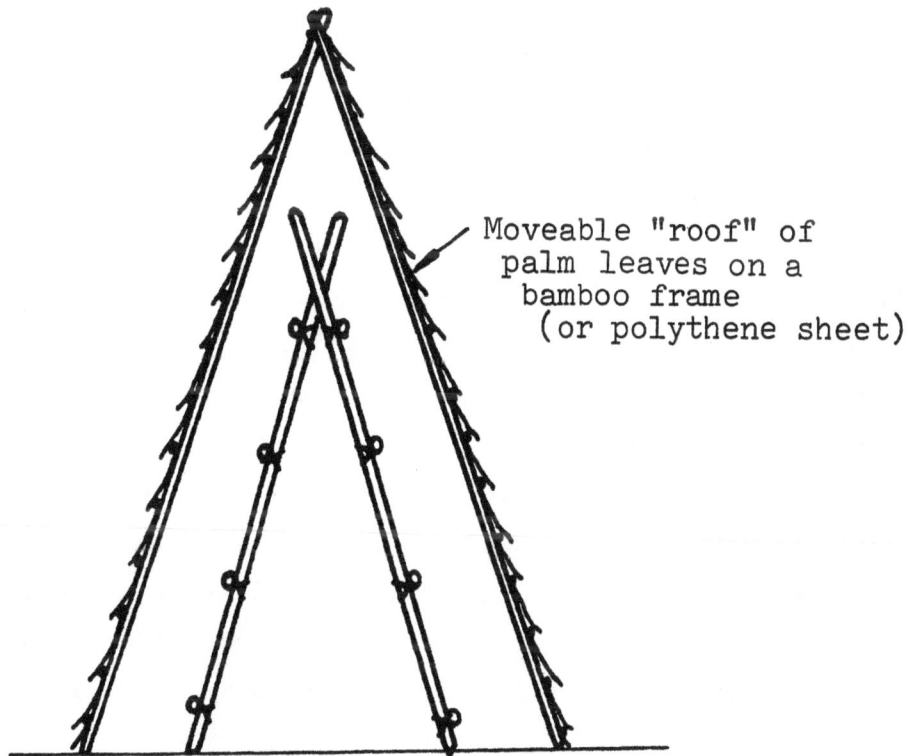

Moveable "roof" of
palm leaves on a
bamboo frame
(or polythene sheet)

Figure II.3

Drying rack inside a moveable "roof"

Maize grain, shelled directly after harvest, may be spread on the ground
and sun-dried to a safe moisture content. Drying under these conditions can
be accelerated by placing the grain on a black surface, such as a black
plastic sheeting (see figure II.4). Two posts are knocked into the ground on
either side of the plastic sheet and a rope or bamboo pole is suspended
between them. The grain is spread over the sheet with the two ends of the
latter left uncovered. In the event of a rain shower, the two ends can be
lifted and pegged to the rope, thus forming a simple tent over the grain.
This type of drying system requires frequent mixing of the grain in order to
prevent over-heating of the latter.

III.2.2 Solar dryers

Solar dryers are often advocated for grain drying (Soza, 1979). These
dryers are particularly useful in areas with high rainfall and humidity during
the post-harvest period. They make use of the limited amount of sunshine for
a rapid drying of the crop.

Figure II.5 shows one type of solar dryers used in large-scale
operations. These dryers are made of three parts:

- a solar collector to heat the drying air;
- a drying bed system which is not exposed to the sun; and
- a fan or similar device for providing a flow of air through the
  solar collector and drying bed.

These dryers were originally designed to improve the drying rate of maize
grain contained in conventional bins. The roof and sun-facing wall of the bin
are converted into a solar collector by painting them black to enhance the
absorption of solar radiation. An air duct is formed by fixing wooden panels
under the roof and inside the sun-facing wall. The heat absorbed by the black
surface is transmitted by conduction to the air within the duct thus raising
its temperature. A fan at the base of the air duct draws the warm air from
the duct, and forces it through the drying bed. After passing through the
grain, the drying air exhausts through the chimney in the roof.

Other types of solar dryers have been developed for the drying of maize
and other grains. Some of these dryers are described in detail in a
publication by VITA (1977).

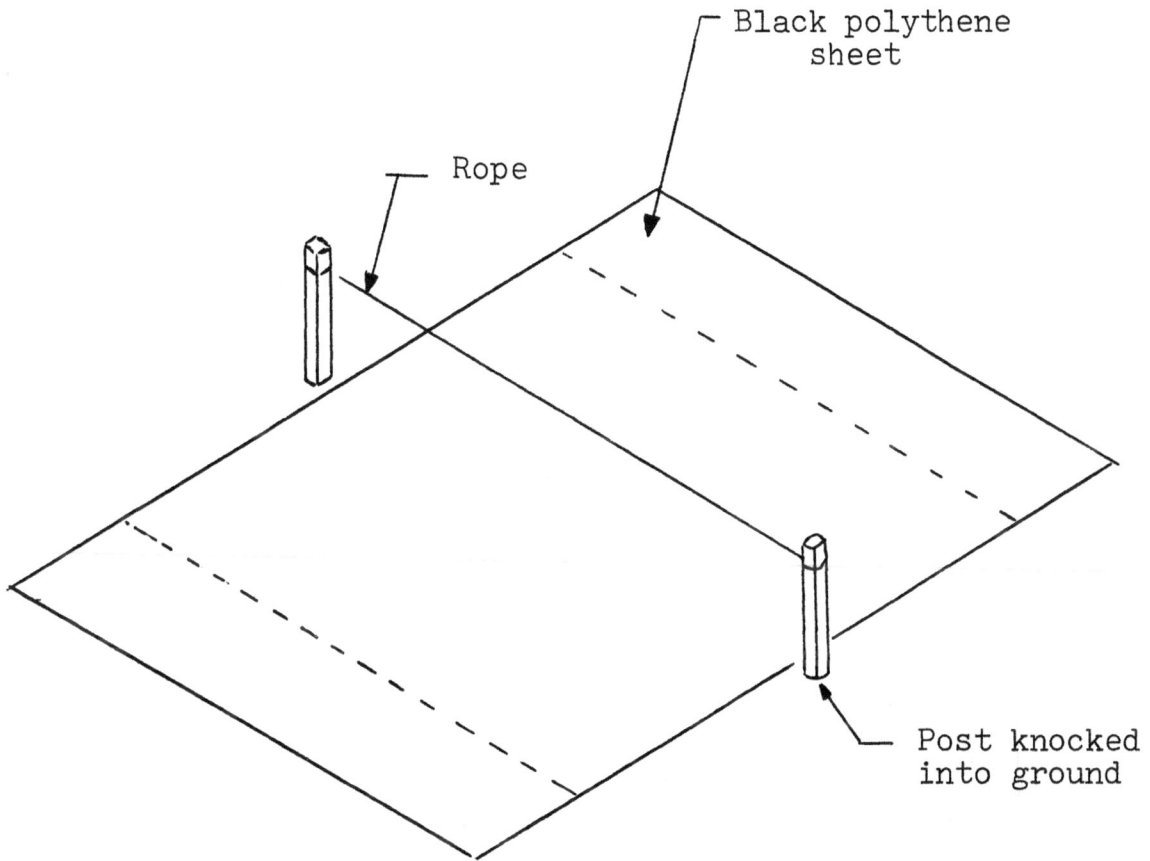

Figure II.4

Improved sun drying of shelled maize grain
Source: O'Kelly (1979)

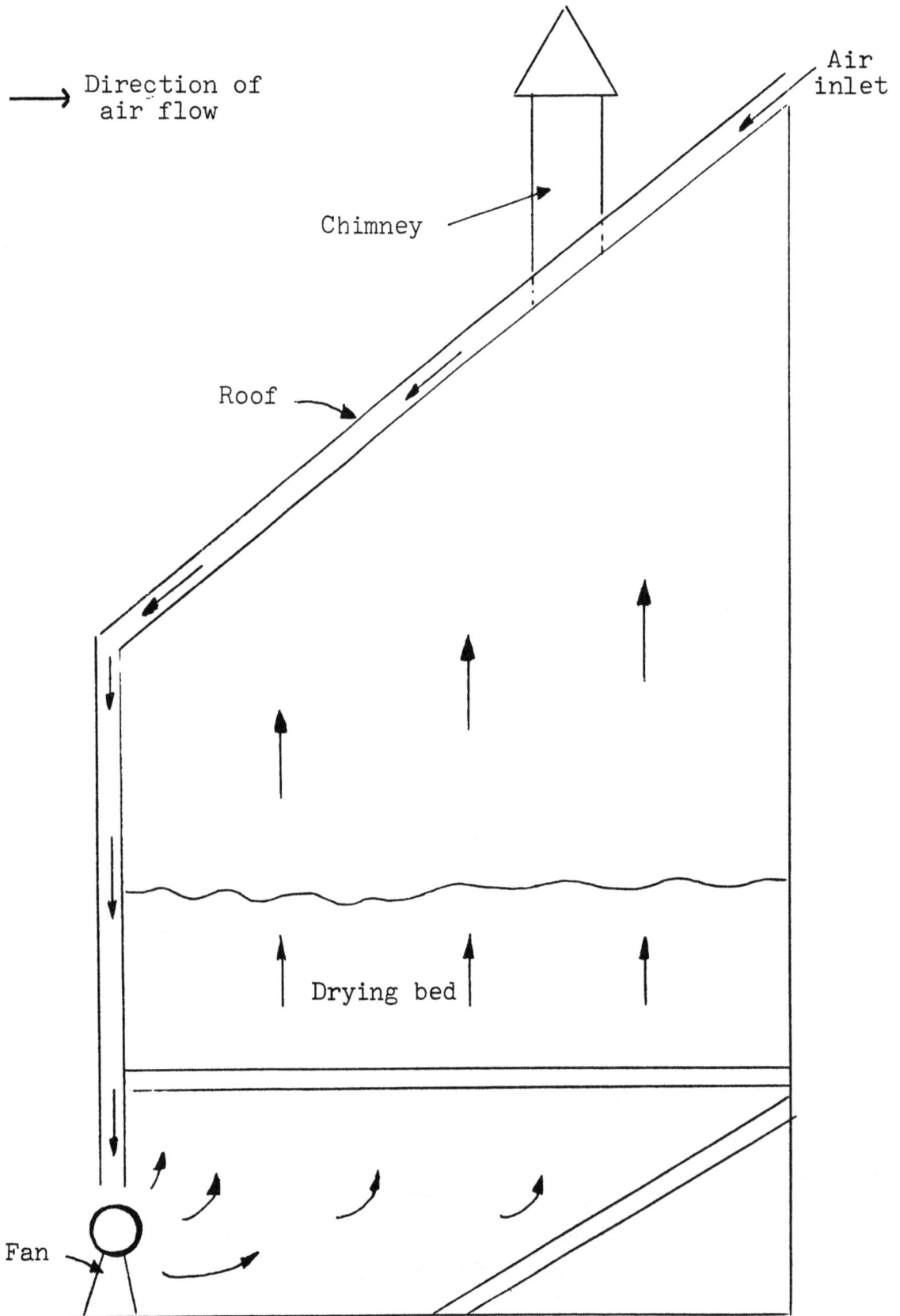

Figure II.5

A forced convection solar dryer
Source: Buelow (1961)

III.2.3 <u>Artificial drying</u>

Despite the fact that solar energy is often utilised for the drying of maize (e.g. on the cob or as shelled grain), unfavourable climatic conditions will often prevent drying to a sufficiently low moisture content for safe storage. Artificial drying, with or without a supplementary heat source, is therefore necessary.

Considerable attention has been paid to the design of simple dryers for the subsistence level farmer. One such dryer, which can be constructed from easily available materials, is illustrated in figure II.6. It is essentially a thick walled clay cylinder with a thatched roof. At mid height, a clay floor, supported by clay pillars, serves as a "heat exchanger" and as a means of supporting the material to be dried. Beneath this 'heat exchanger' is the firing chamber in which wood or agricultural waste is burned to generate the heat required for the drying process. Small smoke outlets in the firing chamber can be closed to reduce combustion. Air inlets in the upper drying chamber allow air to be drawn into the dryer and be heated as it flows through the 'heat exchanger'. The heated air exhausts through the thatch after passing through the rack of drying maize.

An alternative to the natural convection system described above consists in using fan-powered, forced convection systems with or without a supplementary heat source. The fans can be driven by electricity, diesel, petrol or any other rotative power source. Grain may be dried loose in the field in a system similar to that illustrated in figure II.7. Warm air from a fan and heater unit is channelled via a flexible duct into the air space beneath a drying frame covered with hessian sacks. The grain is piled over the frame to a depth of 1.8 m. It is contained within a barrier of grain-filled sacks approximately 1.2 m away from the sides and one end of the drying frame. This system can also be used for drying grain in sacks which should not be more than three-quarters full (see figure II.8). For grain with a moisture content greater than 20 per cent, the sacks should be laid two rows wide around the frame and up to five layers deep over it. With grain at a moisture content of less than 20 per cent, loading can be continued to a greater depth. Since these types of dryers are exposed, a large waterproof cover should be available for protection in the event of rain.

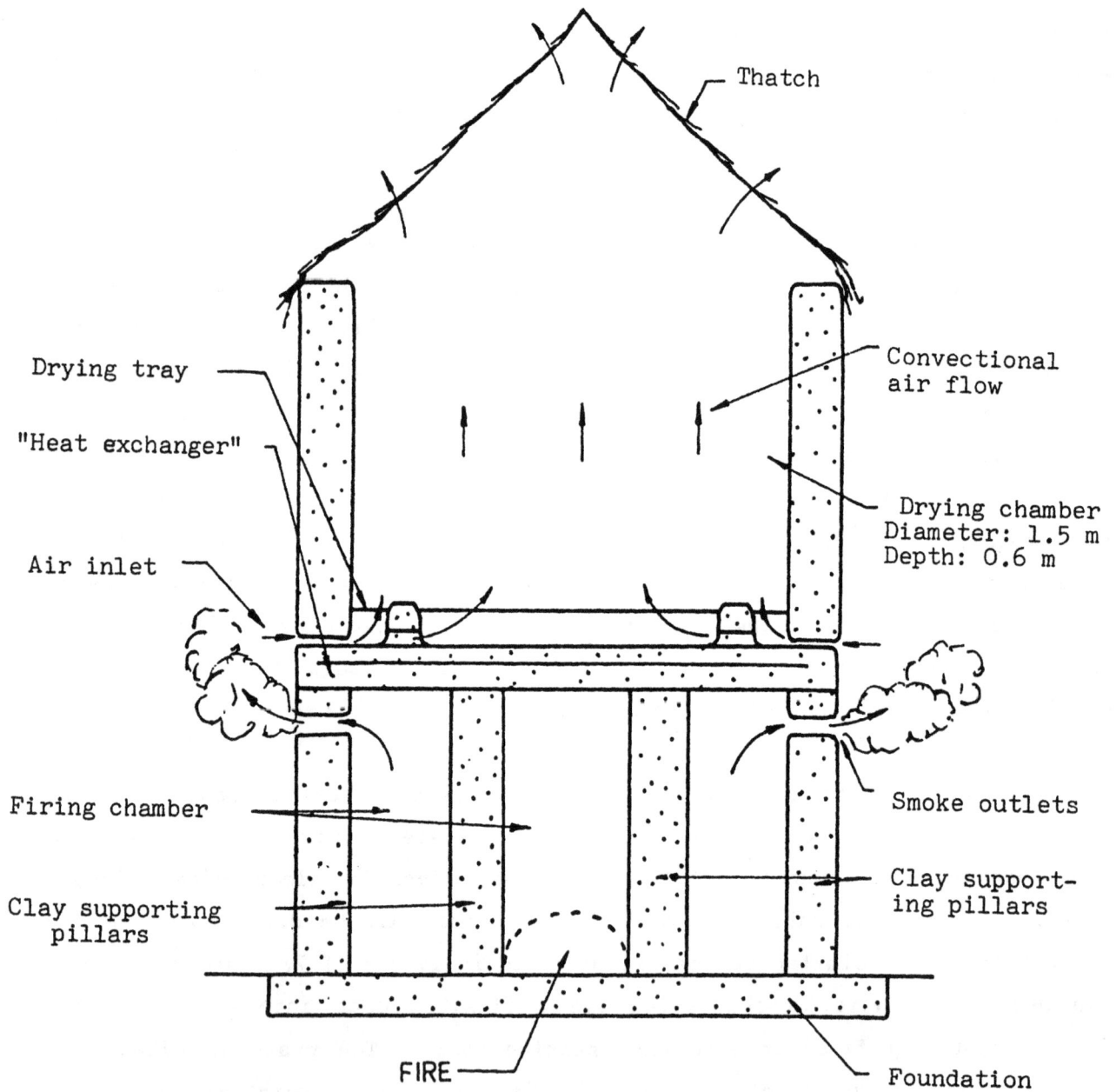

Figure II.6

Fuel fired dryer

Source : O'Kelly (1979)

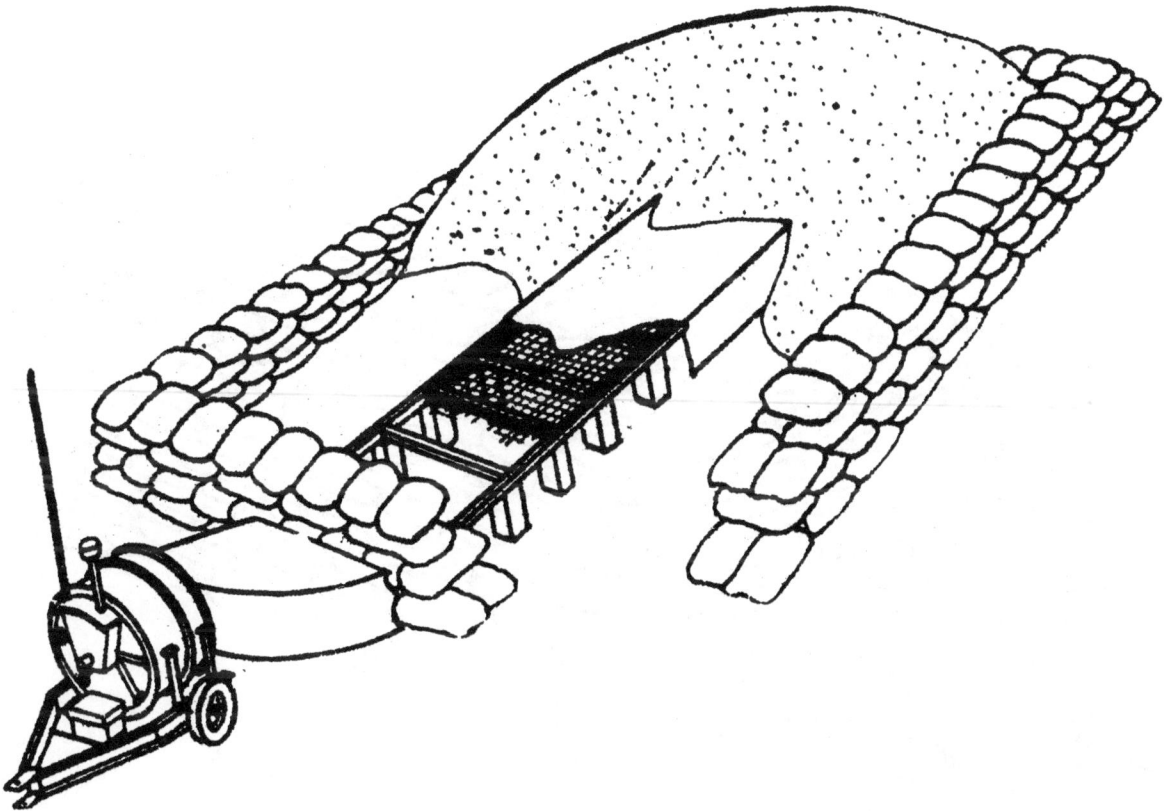

Figure II.7

Ventilated floor drying of loose
shelled grain

Source: R.A. Lister Ltd.

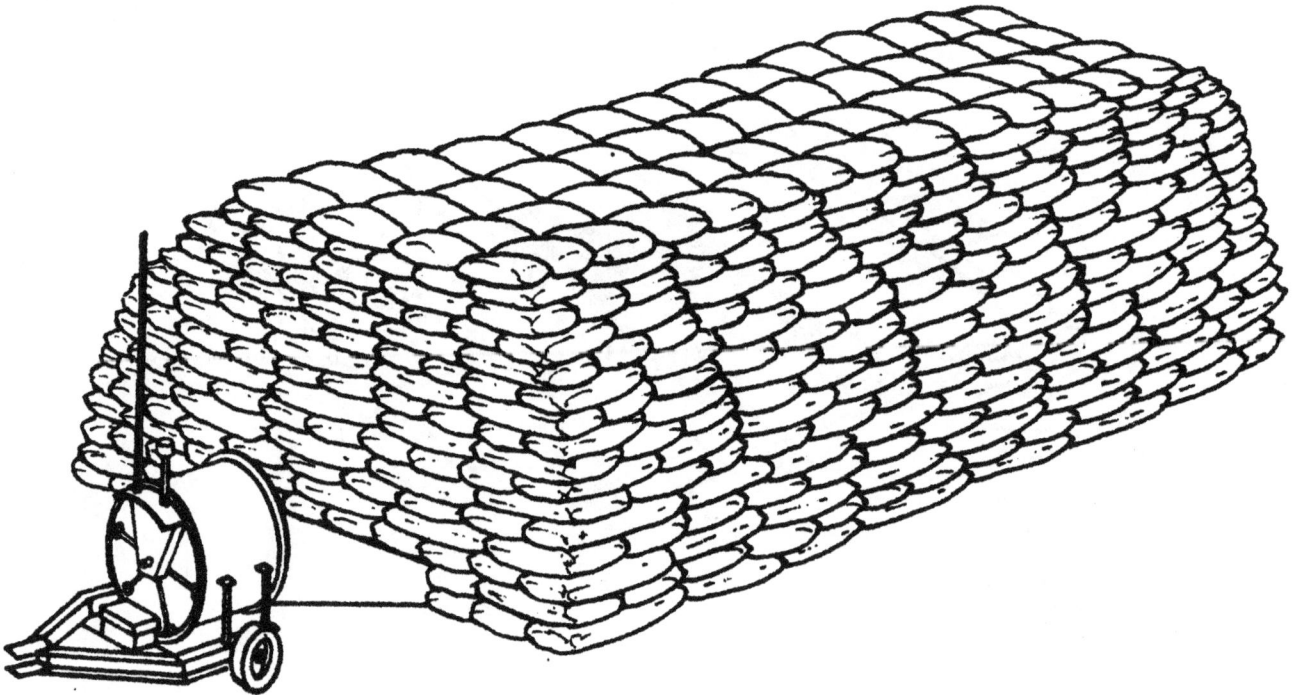

Figure II.8

Ventilated floor drying of sacked grain
Source : R.A. Lister Ltd.

IV. TEMPORARY STORAGE OF DRIED MAIZE

IV.1 Moisture content and safety from moulds

To minimise the risk of mould damage, maize grain should have a moisture content below 13 to 13.5 per cent (wet weight basis) for storage in bags, or below 12 to 12.5 per cent for bulk storage in bins or silos. In general, the drier the grain, the lower the risk for grain damage. However, drying below 12 per cent may lead to increased breakage of grains during handling, and will also require extra conditioning of the grain for milling. Maize grain on the cob can be stored, safe from mould damage, in open-sided cribs. In this case, the moisture content need not be less than 13.5 per cent. Cobs can be loaded into the crib at high moisture content and allowed to dry by natural aeration. Cribs should be of reduced width if used for drying in humid climates.

IV.2 Prevention against insect damage

Dry maize grain, clean and free from evident damage and insect infestation at the time of delivery, will not normally suffer severe insect damage if stored for 3 to 4 months or less. High grain damage levels, including broken grains, increase the susceptibility to insects and moulds, high-yielding varieties being commonly more susceptible to insect pests than traditional varieties. If the initial percentage of insect-infested kernels is about one per cent, the level of damage in 3 to 4 months storage at 20-25°C is likely to rise, at least to about 20 per cent. Such a level represents a 2 to 3 per cent loss of dry weight. At higher storage temperatures (up to about 30°C), the loss is increased, and at mean temperatures of 30°C, it could be more than doubled.

Grain with evident infestation should be disinfested by fumigation with an approved insecticidal gas (e.g. phosphine or methyl bromide, which are approved for this purpose in most countries), or by spraying the grain itself with a locally approved insecticidal grain protectant. A suitable dilute powder formulation can be used instead, and applied by mechanical admixing, if a suitable spray formulation is unavailable or inappropriate to the adopted grain handling system. In general, sprays are preferable for grains handled in bulk, since they can usually be applied to the grain flow at a convenient point in the grain conveying system. Grain in bags can often be more easily treated by shovel-mixing a powder with the grain from each lot of 10 to 20 bags..

IV.3 Prevention against rodent and bird damage

Prevention against bird or rodent damage requires scrupulous attention to site hygiene and the maintenance of effective screens or other barriers to prevent the entry of these pests into the store. Supplementary measures may be needed from time to time to eradicate rodents. Poison-baiting with anti-coagulant rodenticides is the preferred method. Poison-baiting and other techniques for bird control are not recommended and should not be necessary if stores are effectively screened against birds.

IV.4 Storage systems

The common modes of storage and the systematic procedures necessary for their success are indicated in figures II.9 and II.10. Whatever the procedure adopted, management should always place great emphasis on the need for high standards of store maintenance, store and site hygiene, quality control at intake for storage, etc. The systems described in this section constitute a few examples of a large number of grain storage systems available for small-scale operations. Detailed information on other systems may be obtained from the ILO memorandum on grain storage or from a publication by VITA (1977).

IV.4.1 Cribs for storing maize grain
delivered on the cob

The general structure of a crib for storing wet maize on the cob is shown in figure II.9 The sides are made of wire mesh, or of any conventional material, such as loose-woven wattles, that does not obstruct the air flow more than the maize cobs themselves. Natural aeration will dry the cobs slowly but safely whether or not the cob sheaths are removed. Where the climate is generally dry after harvest, the crib may be at least 2 m wide. In humid areas, the width should be reduced to 1 m at most, and possibly to 60 cm.

For protection against rain, the roof should be as water-tight as possible. On the other hand, intermittent wetting of the sides of the cob mass will not seriously impede drying unless the wetting is prolonged and excessive. Extensive roof overhang at the sides is therefore usually unnecessary, and has been shown to reduce the drying rate.

Figure II.9

Crib for storing maize grain delivered
on the cob

The posts and post-holes can, if necessary, be soaked with a persistent insecticide to stop termite damage. Rat damage is prevented by building the base of the crib at least 1 m over the ground. Rat-guards, made of sheet metal cones, may also be mounted on the posts in order to prevent rats from climbing up the crib. These guards should stand out at least 25 cm from the post.

If birds represent a serious problem, additional screening should be used to keep them out. The mesh size of the screen should not exceed 2 cm.

To reduce insect infestation, the crib and the ground beneath it should be thoroughly cleaned before loading. Furthermore, any grain residues that cannot be of immediate use should be burnt. Spray treatment after cleaning, using an approved semi-persistent contact insecticide, is a useful further measure. If the cobs are to be stored longer than 3 to 4 months, they should be treated, layer by layer, as the crib is loaded, with a locally approved grain protectant spray or powder. Alternatively, and more effectively, the grain can be protected by shelling the cobs as soon as they are dry (13 per cent moisture content), and admixing the grain protectant more thoroughly. This permits the effective use of substantially reduced dosage rates for most insecticides.

IV.4.2 <u>Storing bagged dry maize grain</u>
<u>in stacks under cover</u>

The essential features of a good bag stack are shown in figure II.10. A store can hold one very large stack or several smaller stacks. Storage capacity is greatest with a single stack. However, the use of a single stack does not facilitate pest control and decreases its effectiveness.

A covered store can be made rodent-proof and bird-proof. A walled building, when rodent-proofed, facilitates the catching or poisoning of rodents which may gain entry when the doors are open. In the absence of walls, extensive mesh screening may be needed.

Each stack of grain should be bordered by a clear gangway, and an access to the top of the stack should be provided. This is essential for inspection and pest control procedures. If necessary, ladders should be provided for access to the top of the stacks.

a) <u>STORE PLAN</u>

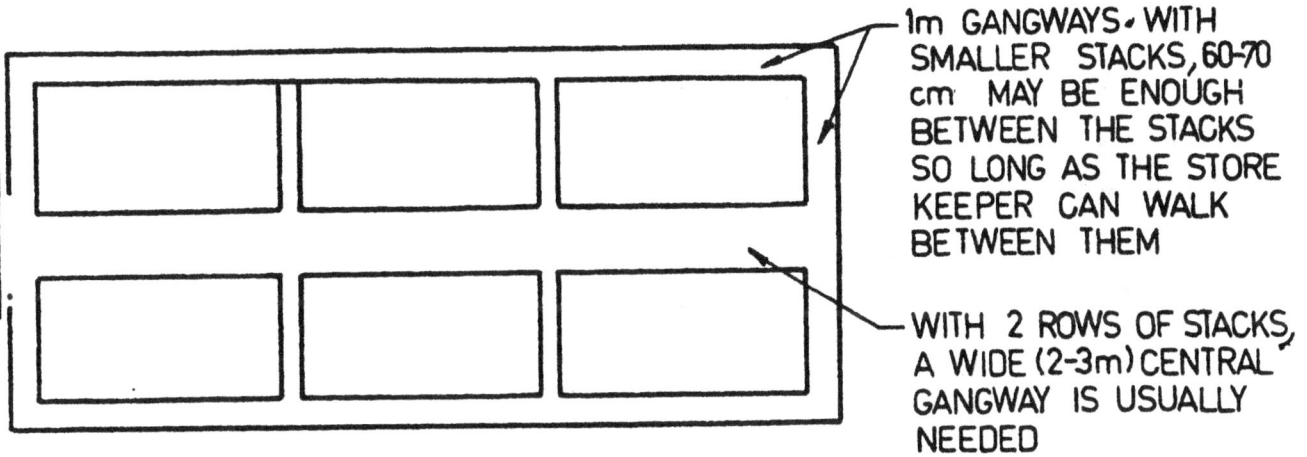

1m GANGWAYS. WITH SMALLER STACKS, 60-70 cm MAY BE ENOUGH BETWEEN THE STACKS SO LONG AS THE STORE KEEPER CAN WALK BETWEEN THEM

WITH 2 ROWS OF STACKS, A WIDE (2-3m) CENTRAL GANGWAY IS USUALLY NEEDED

b) <u>BAG STACK</u>

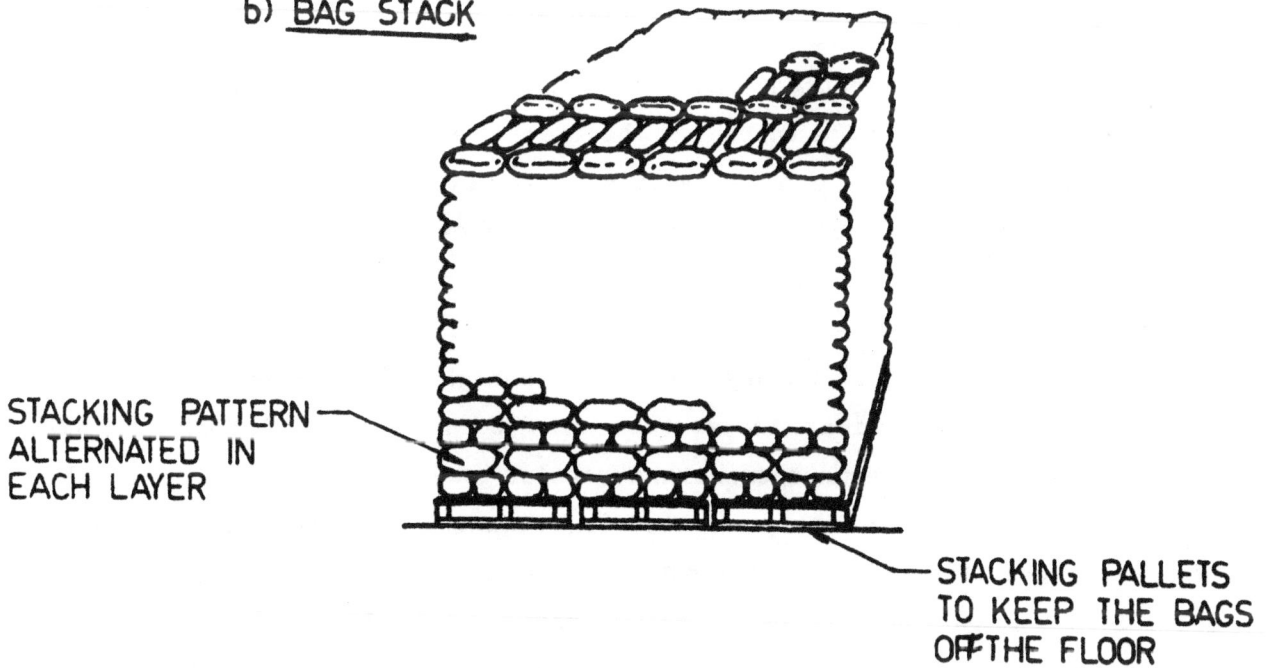

STACKING PATTERN ALTERNATED IN EACH LAYER

STACKING PALLETS TO KEEP THE BAGS OFF THE FLOOR

<u>Figure II.10</u>

Storing bagged dry maize grain in
stacks under cover

In building the stack, the bags need to be kept off the floor by placing them over stacking pallets or regularly spaced poles. The latter may not be necessary if the floor is completely moisture-proof, although they may still be useful in case of minor flood.

The bags should be laid regularly as they are stacked, and should be bonded together by alternating the stacking pattern in each layer as shown in figure II.10. Tight stacking is usually recommended, provided that the grain is dry. Well-built stacks can include 30 to 40 layers of bags in sufficiently high stores.

Stacks can be completely covered with gas-proof sheets for fumigation. The sheets must be in good condition and held down firmly close to the floor by chains or narrow sand bags. For protection against reinfestation after fumigation, the stacks can be sprayed layer-by-layer, as they are built, with a locally approved semi-persistent contact insecticide. Subsequent respraying of the exposed stack surfaces will prolong the protection considerably but not indefinitely. Spraying the exposed surfaces ("capping sprays") without the initial layer-by-layer application is often a waste of time and money. An alternative to spraying treatments consists in leaving the gas-proof sheets in place after fumigation as a physical barrier against insects. With this method, use of insecticides can be restricted to a spray or powder treatment of the floor where the edges of the protective sheets are laid. Excellent, long-term protection can be obtained by this procedure so long as the grain remains dry and cool (i.e. without significant heating above the ambient mean temperature). Caution is necessary, however, in high altitude areas where the daily fluctuations of ambient temperatures may be relatively large. In all situations, good warehouse management and regular inspection for damage to the sheets is essential. Temperature monitoring arrangements are also advisable for the early detection of any infestation problems that could arise due, for example, to a fumigation failure or reinfestation through a torn sheet. The use of transparent, light-weight polythene sheets as permanent fumigation covers is usually less costly and more effective than other methods. Transparency permits a degree of inspection through the sheet, but remote temperature sensing of at least two or three points in the stack is still advisable.

Bag stacks in a store with controllable ventilation can also be effectively protected against reinfestation after fumigation by regular insecticidal space treatments. However, these must be applied sufficiently frequently and at the best time. A daily treatment at dusk is recommended. This method is particularly cost-effective when use is made of insecticides such as dichlorvos. On the other hand, many of the commonly recommended alternatives, especially the natural and synthetic pyrethroids are usually considered rather too costly for daily application in grain stores. A final possibility for stores, that can be made effectively gas-tight and are permanently screened to prevent the entry of insects, is total store fumigation. This method allows a long-term protection of the grain, but requires a type of stores and storage management which are usually impractical where stocks are being moved in and out at frequent intervals.

CHAPTER III

SHELLING

I. GENERAL REVIEW OF SHELLING TECHNIQUES

I.1 Location of shelling

From a purely economic viewpoint, shelling should be carried out at or close to the maize-growing area, either as part of the harvesting operation (e.g. use of a suitably equipped combined harvester), or at a nearby production unit. The main reason for locating the shelling operation close to the growing area is the large difference in weight and volume between the unshelled cob and the shelled grain. Otherwise, the shelling of maize in mills located far away from the point of harvest will considerably increase transport costs. However, there are many occasions where maize is shelled outside the growing area. For example, a large-scale mill, located in an urban centre, may use its own shelling equipment in order to ensure a standard quality of grain (e.g. grain with an appropriate moisture content and free from moulds, contaminants, stones, etc.) and sufficient stocks of the raw material.

Depending on circumstances, and notwithstanding the points raised above, maize may be shelled at the following locations:

- in households, using manual methods;
- in custom mills;
- in small-scale merchant mills;
- in large-scale roller mills; and
- in the field, as part of the harvesting operation.

Since this memorandum is mostly concerned with small-scale milling technologies, this chapter will emphasise shelling techniques which are particularly suitable for custom and small-scale merchant mills. However, some information on shelling tools of particular interest to households will also be provided for the benefit of local manufacturers who may promote these tools among rural households.

I.2 General review of shelling techniques

Maize shelling techniques may include the use of the following devices:

- hand-held devices of various designs and outputs;
- small rotary hand shellers;
- free-standing hand shellers;
- small powered/large hand-operated shellers with cleaning and grading facilities; and
- large powered shellers with loading, cleaning, grading and bagging facilities.

Some shellers used in large-scale roller mills may also incorporate a husking mechanism. Outputs from the above shelling devices range from 15 kg per hour for hand-held devices to over four tonnes per hour for large-scale shellers.

I.2.1 Effect of grain moisture content

The shelling of maize grain is extremely difficult to undertake at moisture contents above 25 per cent, and may not be carried out efficiently at moisture contents between 17 and 25 per cent. Shelling maize at these high moisture contents results in high mechanical grain damage, reduced grain stripping efficiency and high power consumption. In general, efficient shelling is best undertaken after the grain has been suitably dried to a moisture content of 13 to 14 per cent. On the other hand, the use of combine harvesters, for the shelling of maize in the field before it is sufficiently dried, can result in high damage rates of the grain of up to 20 per cent (Waelti and Buckle, 1967).

I.2.2 Shelling damage

Appropriately dried maize may still be damaged during shelling (e.g. a fraction of the grain may get surface scratches or break). Such damage will be of little consequence if the maize is to be further processed shortly after shelling (e.g. it may be milled or cooked). On the other hand, shelled maize which must be stored over a long period will deteriorate as a result of attacks by secondary pests or fungal invasion which spread through the damaged surface of the grain. This deterioration will happen even if the grain is sufficiently dry and properly stored.

The extent of shelling damage depends on the shelling technique, the skill of the operator, and the type of maize being shelled. In general, manual shelling methods - with exception of those involving scrapping and beating - and efficient small shellers result in much less grain damage than do large, powered shellers. However, many farmers or mill operators do not use these small shellers efficiently and do produce a high fraction of damaged grain for the following reason. A number of rotary hand shellers are equipped with an adjustable spring which allows an increase in the pressure on the cob in order to improve the shelling action. Since most farmers cannot tolerate the loss of unstripped grain nor the extra effort required to recover it by hand, they tend to adjust the spring to a high pressure setting in order to ensure a complete stripping of the cob. Such a high pressure results in extensive damage of the grain surface, and in the loss of grain if the latter is to be stored over a long period of time.

I.2.3 Factors affecting the choice of shelling technique

Depending on the scale of production, one or more shellers may produce the required output of shelled maize. The most appropriate shelling technique will thus be the one which minimises production costs per unit of output for a given loss rate of damaged grain, quality of output (e.g. extent to which the shelled grain is free from sand, stones, dead pests, etc.) and the quality of raw materials (e.g. moisture content, size and quality of the cobs, the relative softness of the grain endosperm).

When comparing two alternative shelling techniques, one must take into consideration all ancillary operations since some shellers may carry out some or all of these, while other shellers may only carry out the shelling operation. Thus, such a comparison should take into consideration the cost associated with the following ancillary operations:

- removal of leaf sheaths;
- cleaning; .
- sieving; and
- bagging, whenever required.

For a given quality of raw materials, the shelling cost must therefore include the following cost items:

- husking costs;
- cleaning and sieving costs;
- bagging costs;
- loss of revenues due to the presence of damaged grain in case the shelled maize must be stored (i.e., the sale of damaged grain is lower than that of grain in good condition).

The methodology for estimating these cost items with a view to selecting the most appropriate shelling technique is described in Chapter VI.

The choice of shelling technique should take into consideration the following variables:

(i)   the total volume of crop to be processed;
(ii) the effective length of the shelling season
      and of the working day;
(iii) the actual capacity of available manual and
      engine-powered equipment;
(iv) the equipment and labour productivity; and
(v)   the cost of the factors of production, including
      equipment costs, wages, interest rate, energy cost, etc.

Variables (i) and (ii) determine, for a given location, the minimum daily volume of maize which must be processed. The longer the length of the shelling season, and/or the larger the storage capacity, the lower will be the minimum daily volume. Variable (iii) and (iv) determine the number of each type of maize shellers for the processing of this minimum daily volume, taking into consideration the fact that some of the shellers are not equipped for the carrying out of ancillary operations. Finally, given the number of each type of shellers, and the labour, infrastructure and energy requirements for each type, the variables under (v) determine the unit production cost associated with each shelling technique.

It may be noted that the installed capacity of the shelling equipment may be affected by the following factors:

- the availability of sufficiently skilled labour and the
  organisation of the shelling operation as a whole;
- the level of moisture in the crop;
- the size and quality of the cobs;
- the extent to which dehusking is required of the machine; and
- the rate of supply of cobs.

The estimated capacity is often lower than the actual capacity, the difference between the two being generally larger for large-scale equipment than for small-scale equipment. One reason for this larger difference is that it is more difficult and time-consuming to maintain and repair large shellers than small shellers, especially the manual type shellers.

II. SHELLING EQUIPMENT

There are various types of shelling equipment, each type catering to differing scales of production and conditions (e.g. domestic use, co-ownership by small farmers, use by independent farmers on a daily hire basis, ownership by custom or merchant mills). Farmers often use small, hand-cranked, rotary shellers which are usually simple, effective, inexpensive and fairly durable. They are available from many manufacturers from both developing and industrialised countries. A priori, any developing country should be able to manufacture this type of shellers. The latter are of various designs, and made from a small number of iron castings. They utilise a spiked disk to prise out the grain while the cob is held by an adjustable spring-loaded pressure plate. While the capacities of different machines vary, these will at least double the rate of the most productive hand-held devices (e.g. 100 kg of grain per hour).

The larger, free-standing shellers are more productive and convenient, but more expensive. They are often fitted with cleaning and separation devices for the removal of unwanted material. The relatively large size of the maize grain facilitates the use of both cleaning fans for the blowing away of dust and light particles, and of simple reciprocating sieves for the removal of sand, stripped cob centres and broken or undersized grain. Depending on the type of sheller and on the number of operators employed, the capacity of these machines can be four times larger than that of the smaller rotary shellers. This may be explained by the use of low-friction bearings and of simple gearing which result in steady and high operating speeds. The shelling principles are similar to those of the smaller shellers. One main difference is the replacement of the spring-loaded pressure plate by a relatively low speed feed roller which forces cobs into the shelling element. The design and sizing of large shellers allow the use of alternative drive methods, such as small electric or petrol motors. These shellers are suitable for small-scale merchant mills.

The full-sized, diesel or electrically-powered shelling machines, with capacities of several tonnes per hour, represent the normal equipment used in large-scale, fully mechanised situations. A large number of firms produce their own designs for sale through normal agricultural equipment suppliers. No standardised design exists, but most shellers use broadly similar operating principles. These shelling machines are available in a variety of intallations. Use within a mill requires a fixed installation, with associated handling and feeding facilities fitted close by. Mobile installations are also available. They are either wheeled or mounted on tractors. The high rates of throughput require the use of cob loading elevators and bagging equipment. Most shellers utilise a pegged drum, mounted on a horizontal shaft, which rotates at about 700-1,000 rpm. A concave metal screen, with holes approximating to the size of the grain, is located around the drum. It contains the cobs while shelling takes place. A baffle plate restricts the flow of he cobs, and maintains the required shelling pressure. A strong fan discharges the stripped cob centres and other large debris. A second, smaller fan is often used at the grain discharge point for the removal of the remaining dust and finer particles. Available information indicates that an average shelled grain output of 900 kg/hr/installed kWh may be obtained from these shellers.

Dehusking of the cob may be carried out with a special device installed within the shelling machine. The usual method of dehusking is to provide sets of contra-rotating rollers whose projections pull the husk away from the cob. It is possible to shell and husk maize, despite some loss of capacity, within the shelling section itself. However, it is recommended to obtain the advice of individual manufacturers on the practicability of this approach.

The working lifetime of the various components of these machines should be relatively high since no wearing or rubbing parts are employed. Thus, the need for spare parts, such as bearings, drum parts and screws should be relatively low. However, differences in materials of construction do occur and no typical replacement rates can be quoted. No skilled labour is required to run these machines since it is only necessary to manually feed the cobs and dispose of the grain and cob centres. The number of labourers employed has a pronounced influence on the work rate of the machines, particularly that of the smaller units with unmechanised loading.

II.1 Hand-held shelling devices

A number of hand-held devices have been developed with a view to improving labour productivity, reducing the tedium of work and minimising finger soreness. The hourly output of shelled grain achieved by these devices ranges from 8 kg to 15 kg per hour. Figure III.1 shows a number of hand-held devices developed in recent years. Table III.1 provides information on the designers/producers of these devices, the main construction method and skill requirements, their relative production cost and the estimated hourly output. It may be noted that all of these devices may be produced by local workshops using imported or locally available materials.

II.2 Small rotary hand shellers

These shellers are particularly suitable for small-scale production. They are effective and usually quite robust. Their operation is fairly simple, although grain damage may result from inadequate adjustment of the equipment. The turning of the handle forces the cob to rotate against the spikes of a disc which removes the grain. Available designs allow for the separation of the spent cob centre from the grain. This, however, requires that the owner improvise a mounting stand and grain collection facility. Designs from various manufacturers differ in surface finish, quality of alignment and material, and in the thickness of metal. These variations result in different outputs ranging from 14 to over 100 kg per hour. Whilst these shellers are best made in metal, it is possible to produce similar equipment from wooden constructions. They may be manufactured locally by skilled craftsmen and entrepreneurs.

Table III.1

Characteristics of hand-held shelling devices

Sheller type	Source	Description	Construction	Construction skill required	Relative cost (local manufact.)	Capacity kg/hr
Decker	See manufacturers list	See illustr.				15
Morogoro	Prof. A.S. Ramo, University of Dar-es-Salaam, "Appropriate Technology", Vol. 2, No. 1, May 1975	See illustr.	Using saw timber	Fairly high	High	13
Ceneema	"Bloc-notes du monde rural", No. 13, June 1977, B.P. 790, Yaoundé (Cameroon)	See illustr.	Welded construction	Fairly high	Medium	10
TPI	"Rural Technology Guide No. 1", 1977, Tropical Products Institute, London	See illustr.	Wooden carving	Low	Low	10
PVC Pipe	Dr. D.J. Hilton, University of Nairobi, "Appropriate Technology" Vol. 3, No. 2, Aug. 1976	See illustr.	Glued PVC piping	Low	Low	8
SATA	Swiss Association for Technical Assistance, P.O. Box 113, Kathmandu, Nepal	As Decker	Folded sheet	Fairly high	Medium	As Decker

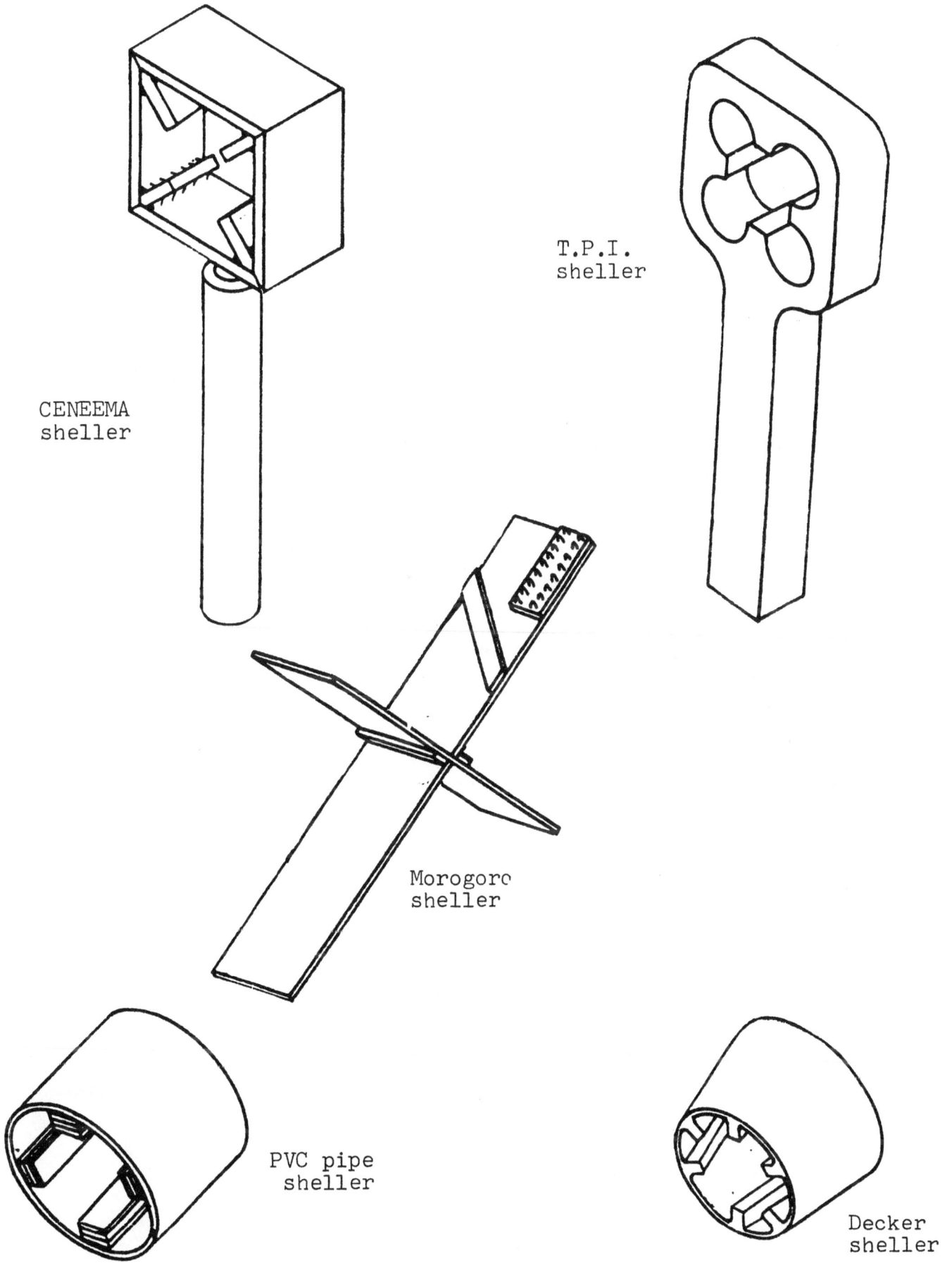

CENEEMA
sheller

T.P.I.
sheller

Morogoro
sheller

PVC pipe
sheller

Decker
sheller

Figure III.1

HAND-HELD SHELLING DEVICES

Manufacturers from both developing and industrialised countries produce a large range of rotary hand shellers. Some of these may be adjusted for different sizes of cobs, and can be mounted on various frames. Plates III.1 to III.6 provide a brief description of some of these shellers for illustrative purposes. The quoted prices of some of the shellers are purely indicative. The reader may also wish to obtain information on other shellers from manufacturers mentioned in Appendix I.

II.3 Free-standing hand shellers

The free-standing hand shellers are relatively large and complete shelling machines. They are substantially more expensive than the smaller rotary shellers, but much more productive and easier to operate. The method of grain removal is similar to that for the small shellers, but includes some modifications to improve the capacity of the machine. (e.g. use of a flywheel and of mechanical cob feed rolls). These shellers are invariably fitted with a simple grain cleaning screen or winnowing fan, and have a positive separation of grain and spent-cob. They may be operated in a variety of ways for capacities ranging from 40 kg/hr (single hand operation) to 300 kg/hr (using a small motor and two operators). The design of these shellers varies from one manufacturer to another.

These shellers are particularly suitable for small-scale merchant mills. Plates III.7 to III.9 show two types of free standing shellers.

II.4 Large powered shellers

General information on large powered shellers has already been provided earlier in this chapter. No further technical details are provided in this section since these shellers are not generally suitable for small-scale shelling. Plates III.10 to III.13 describe some of these shellers manufactured in both developing and industrialised countries.

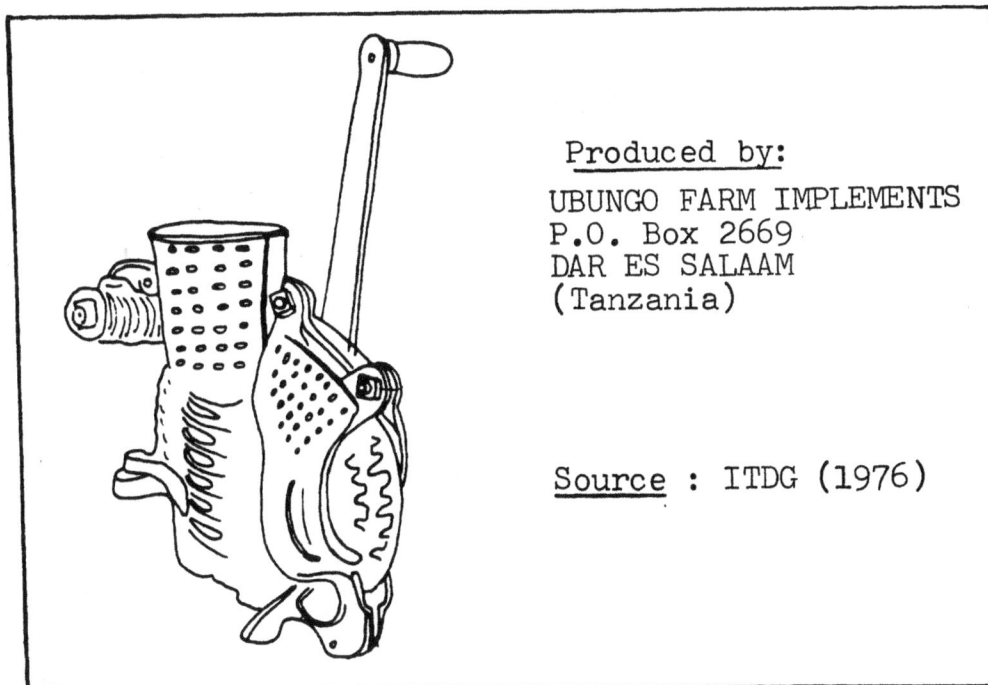

Produced by:

UBUNGO FARM IMPLEMENTS
P.O. Box 2669
DAR ES SALAAM
(Tanzania)

Source : ITDG (1976)

Plate III.1

Maize sheller

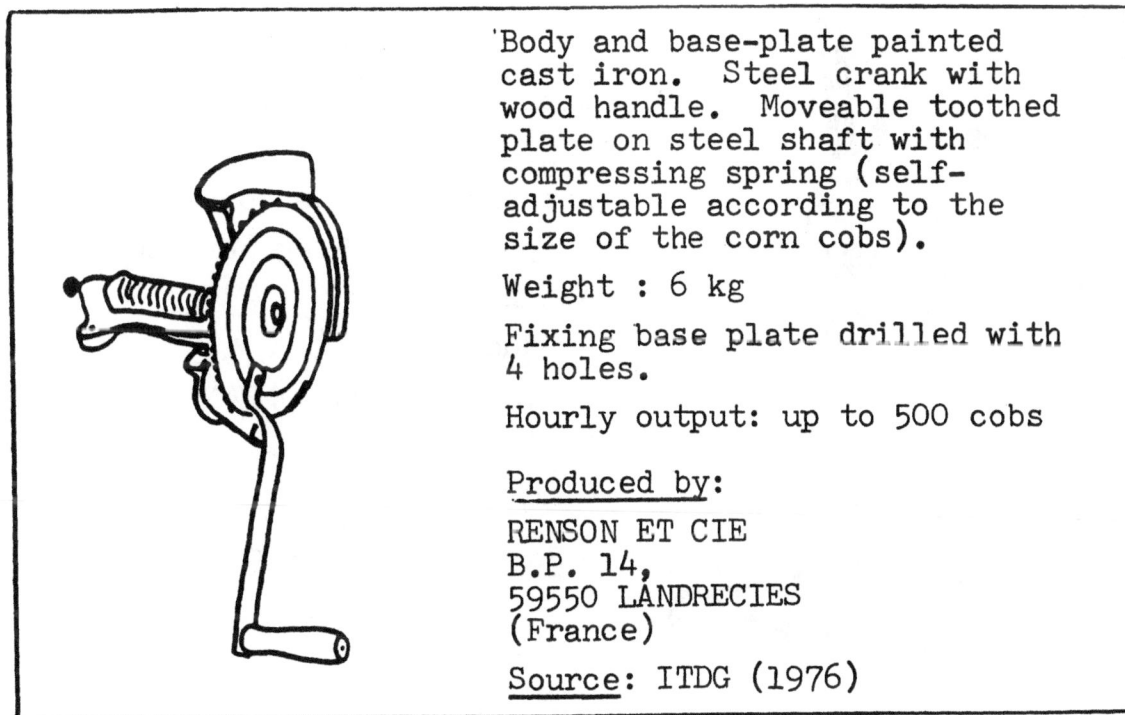

'Body and base-plate painted
cast iron. Steel crank with
wood handle. Moveable toothed
plate on steel shaft with
compressing spring (self-
adjustable according to the
size of the corn cobs).

Weight : 6 kg

Fixing base plate drilled with
4 holes.

Hourly output: up to 500 cobs

Produced by:

RENSON ET CIE
B.P. 14,
59550 LANDRECIES
(France)

Source: ITDG (1976)

Plate III.2

Hand maize sheller

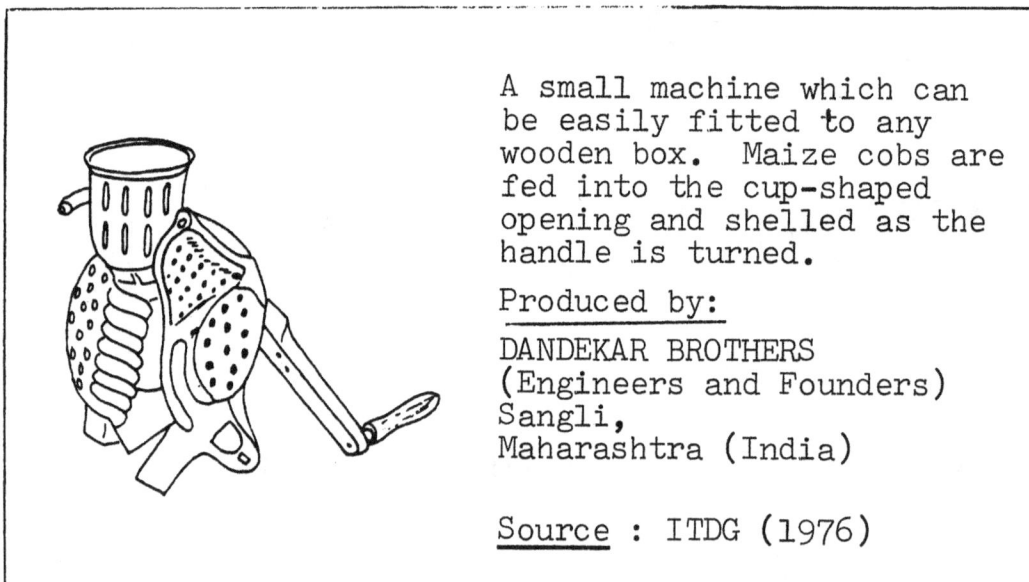

A small machine which can
be easily fitted to any
wooden box. Maize cobs are
fed into the cup-shaped
opening and shelled as the
handle is turned.

Produced by:

DANDEKAR BROTHERS
(Engineers and Founders)
Sangli,
Maharashtra (India)

Source : ITDG (1976)

Plate III.3

Hand maize sheller

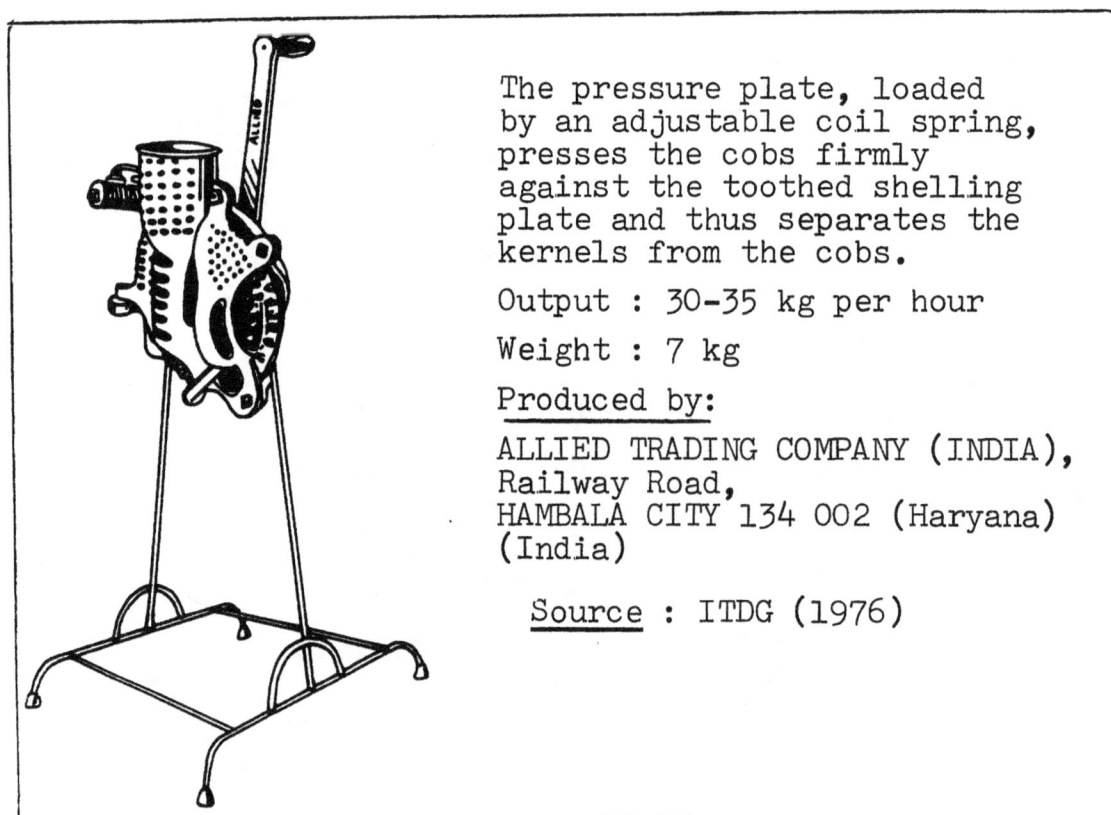

The pressure plate, loaded
by an adjustable coil spring,
presses the cobs firmly
against the toothed shelling
plate and thus separates the
kernels from the cobs.

Output : 30-35 kg per hour

Weight : 7 kg

Produced by:

ALLIED TRADING COMPANY (INDIA),
Railway Road,
HAMBALA CITY 134 002 (Haryana)
(India)

Source : ITDG (1976)

Plate III.4

Allied maize sheller

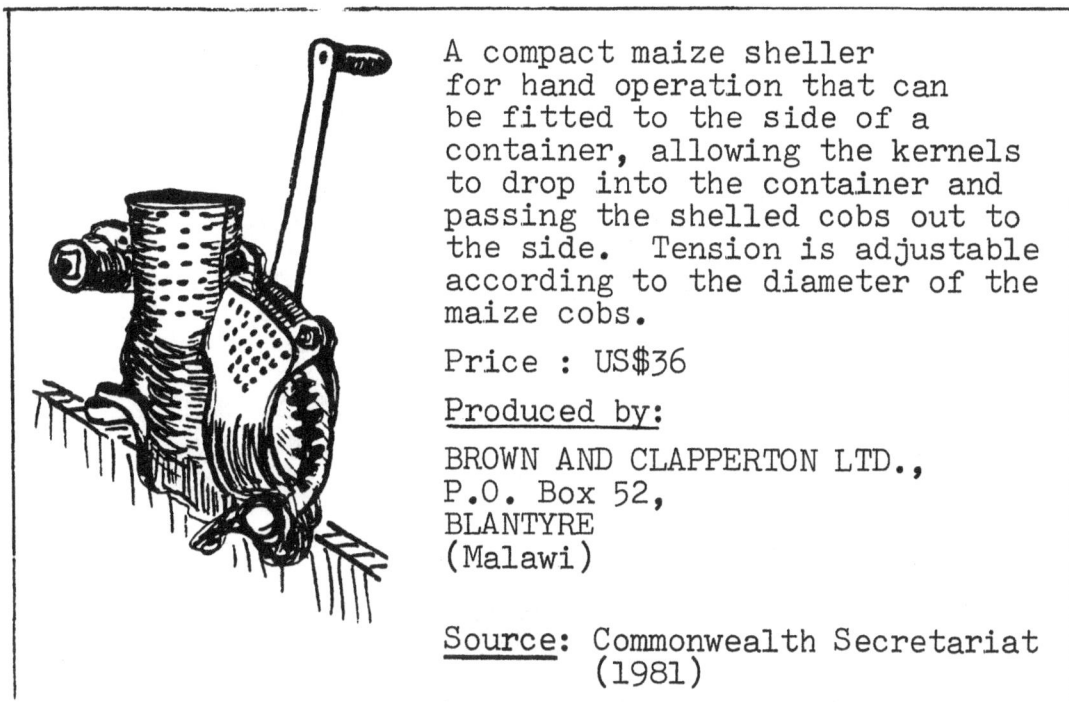

A compact maize sheller
for hand operation that can
be fitted to the side of a
container, allowing the kernels
to drop into the container and
passing the shelled cobs out to
the side. Tension is adjustable
according to the diameter of the
maize cobs.

Price : US$36

Produced by:

BROWN AND CLAPPERTON LTD.,
P.O. Box 52,
BLANTYRE
(Malawi)

Source: Commonwealth Secretariat
(1981)

Plate III.5

Hand maize sheller

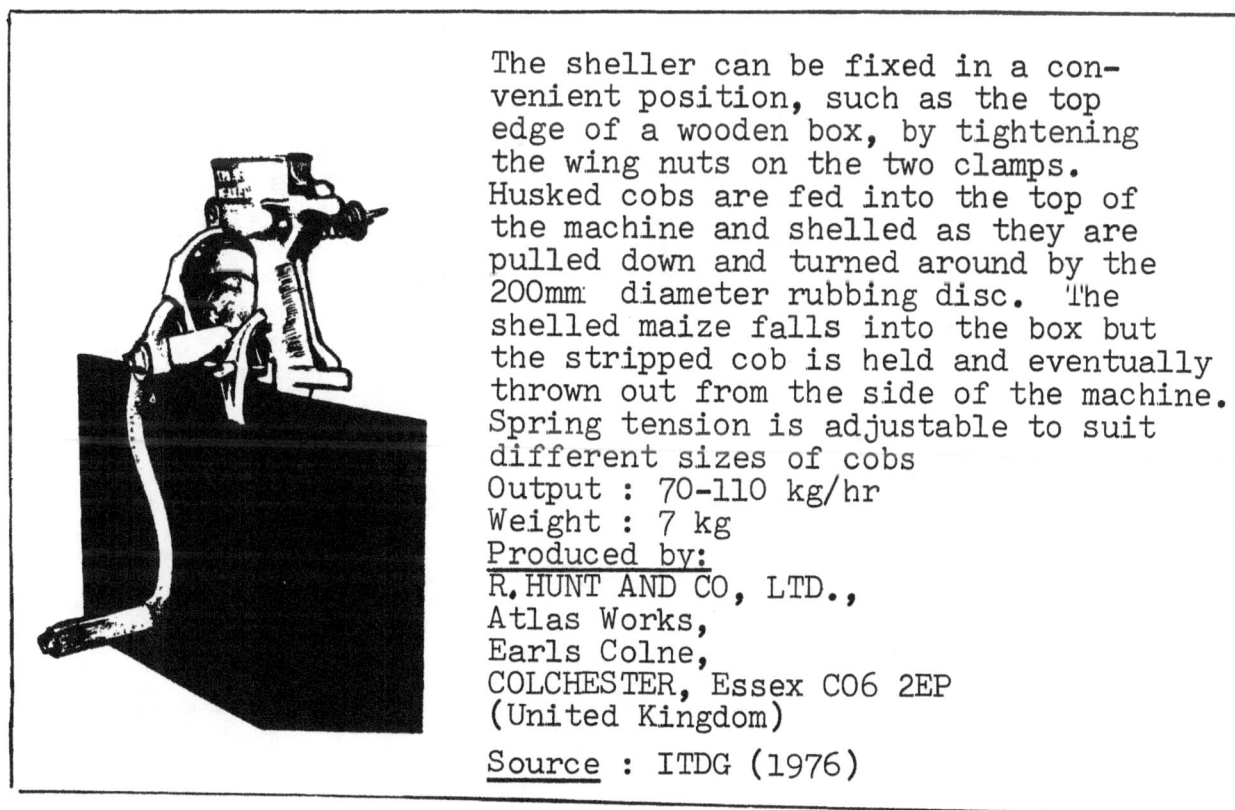

The sheller can be fixed in a convenient position, such as the top
edge of a wooden box, by tightening
the wing nuts on the two clamps.
Husked cobs are fed into the top of
the machine and shelled as they are
pulled down and turned around by the
200mm diameter rubbing disc. The
shelled maize falls into the box but
the stripped cob is held and eventually
thrown out from the side of the machine.
Spring tension is adjustable to suit
different sizes of cobs
Output : 70-110 kg/hr
Weight : 7 kg
Produced by:
R. HUNT AND CO, LTD.,
Atlas Works,
Earls Colne,
COLCHESTER, Essex CO6 2EP
(United Kingdom)
Source : ITDG (1976)

Plate III.6

Atlas maize sheller

Husked cobs are hand fed from a hopper into two feed holes which have adjustable rubbing padds for shelling small or large cobs. A cleaning fan blows light refuse out of the front of the machine, from which shelled cobs are ejected. This sheller may be supplied in hand turned and engine-powered forms, as well as in the pedal powered version illustrated.
Output: 750-900 kg/hr;
Weight : 95 kg
Produced by : RAMSOMES, SIMS
    AND JEFFERIES LTD.
    Ipswich 1P3 9QG,
    United Kingdom

Plate III.7

Cobmaster sheller with pedal arrangement

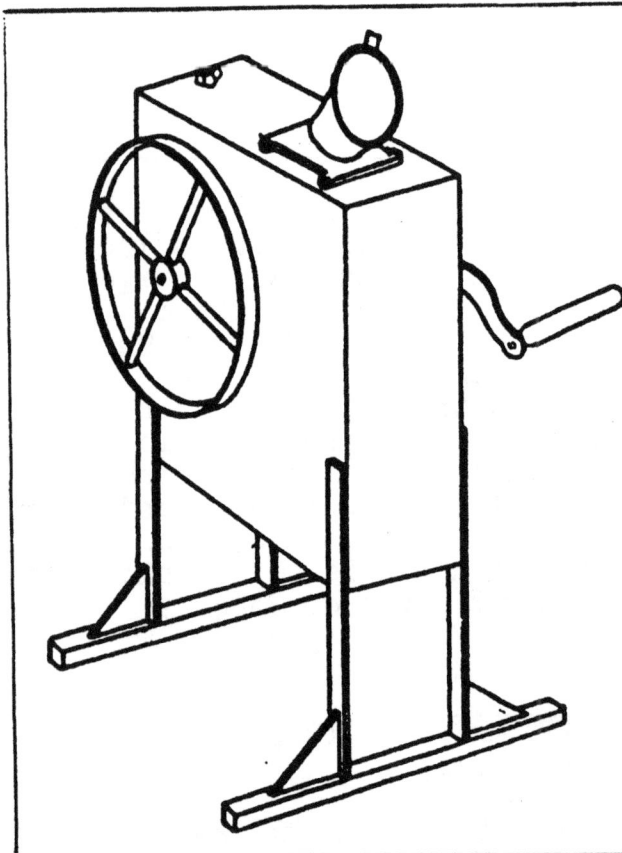

Two models of this sheller are available: a small (manual) model whose capacity is 80-100 kg grain/hour and a large model (manual or motor-driven) whose capacity is 150-200 kg/hour
Manufactured by:SISMAR
B.P. 3214, DAKAR (Senegal)
Price: (as at 1.1.80)
- Small manual : US$33
- Large manual : US$303
- Large motor driven: US$892

Plate III.8

Free-standing hand-sheller

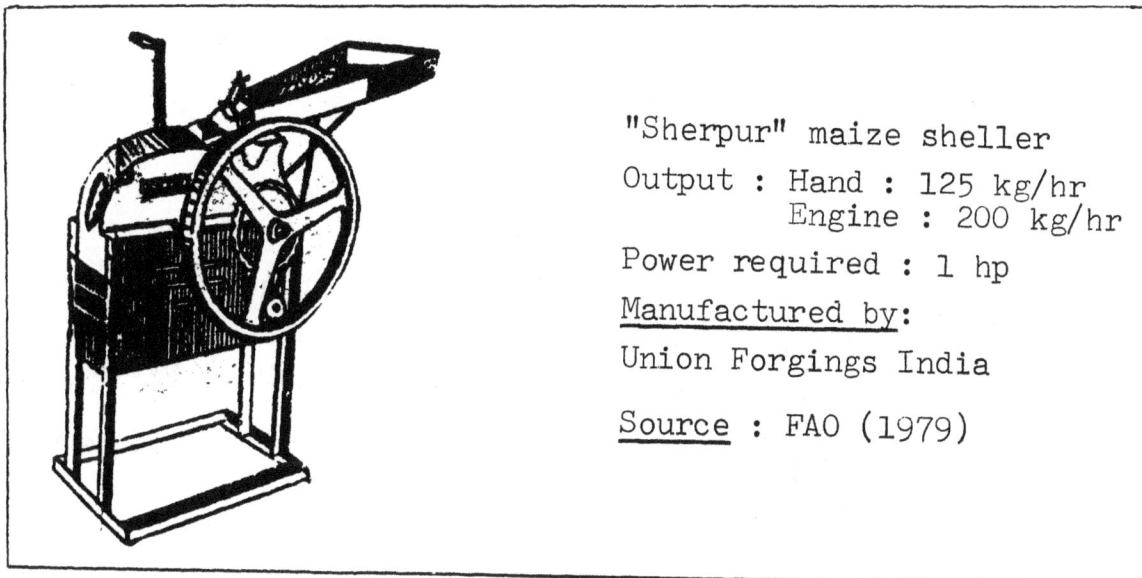

"Sherpur" maize sheller
Output : Hand : 125 kg/hr
         Engine : 200 kg/hr
Power required : 1 hp
Manufactured by:
Union Forgings India

Source : FAO (1979)

Plate III.9

Hand or engine-operated
free-standing maize
sheller

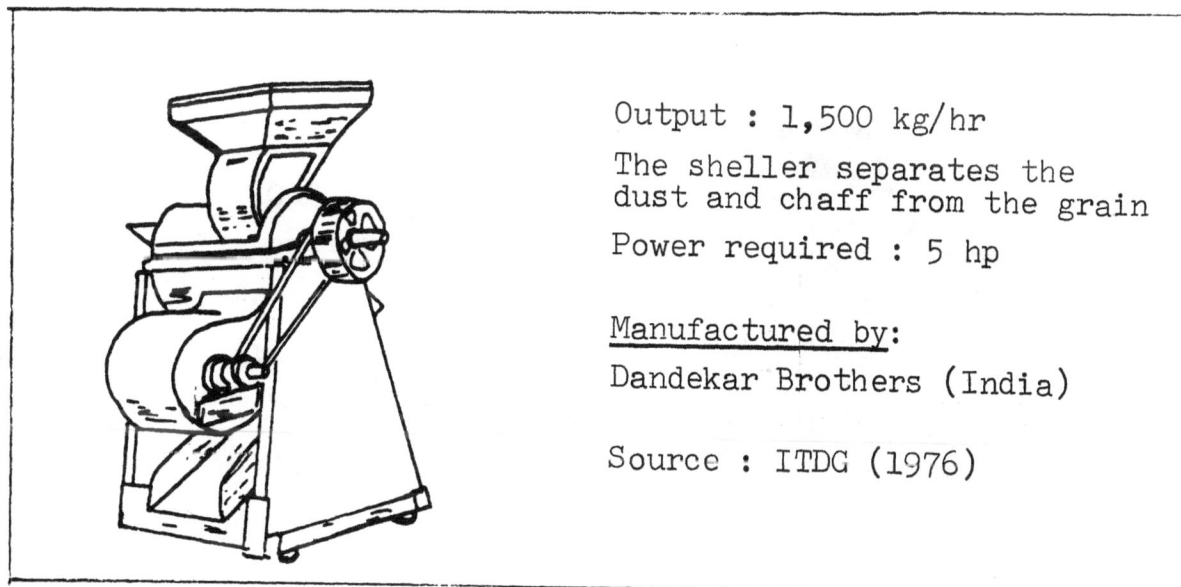

Output : 1,500 kg/hr
The sheller separates the
dust and chaff from the grain
Power required : 5 hp

Manufactured by:
Dandekar Brothers (India)

Source : ITDG (1976)

Plate III.10

Power maize sheller

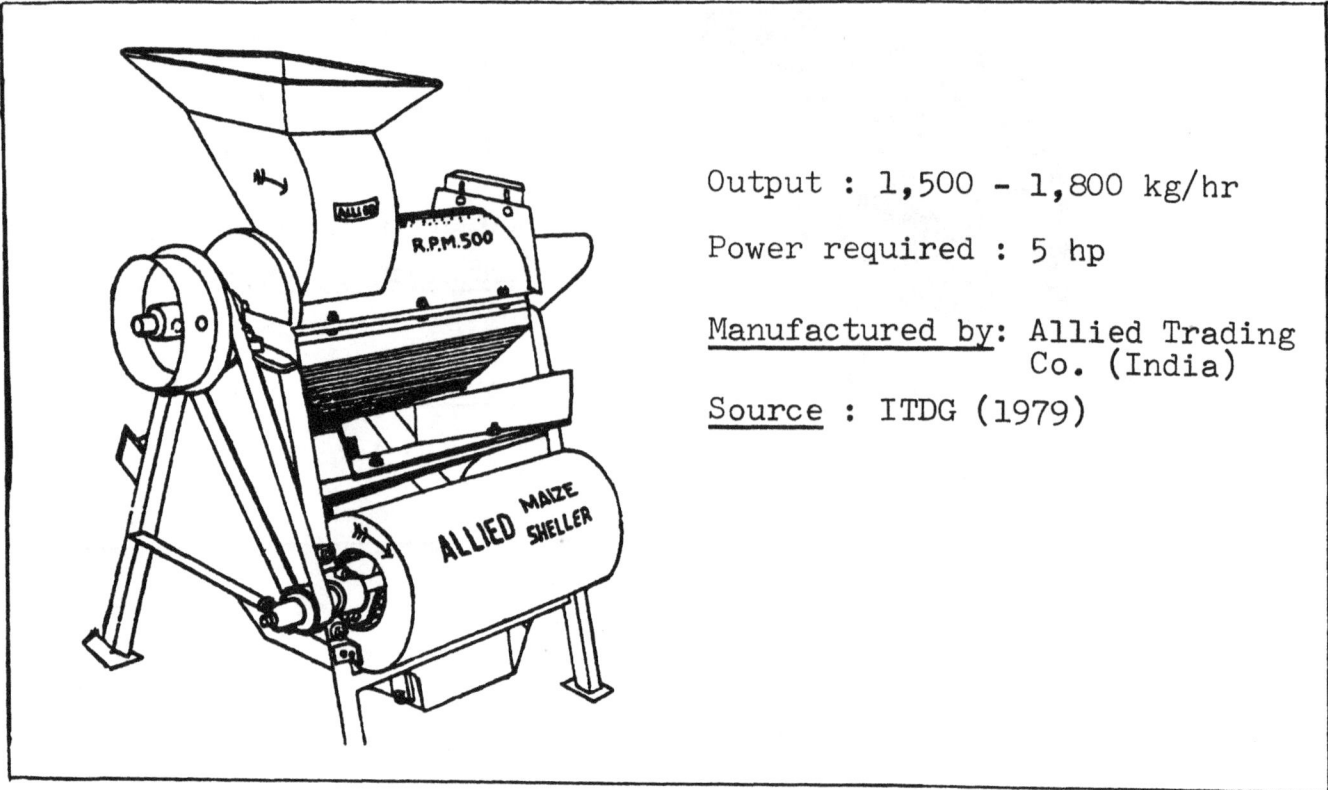

Output : 1,500 - 1,800 kg/hr

Power required : 5 hp

Manufactured by: Allied Trading
                 Co. (India)

Source : ITDG (1979)

Plate III.11

Maize sheller with winnowing fan

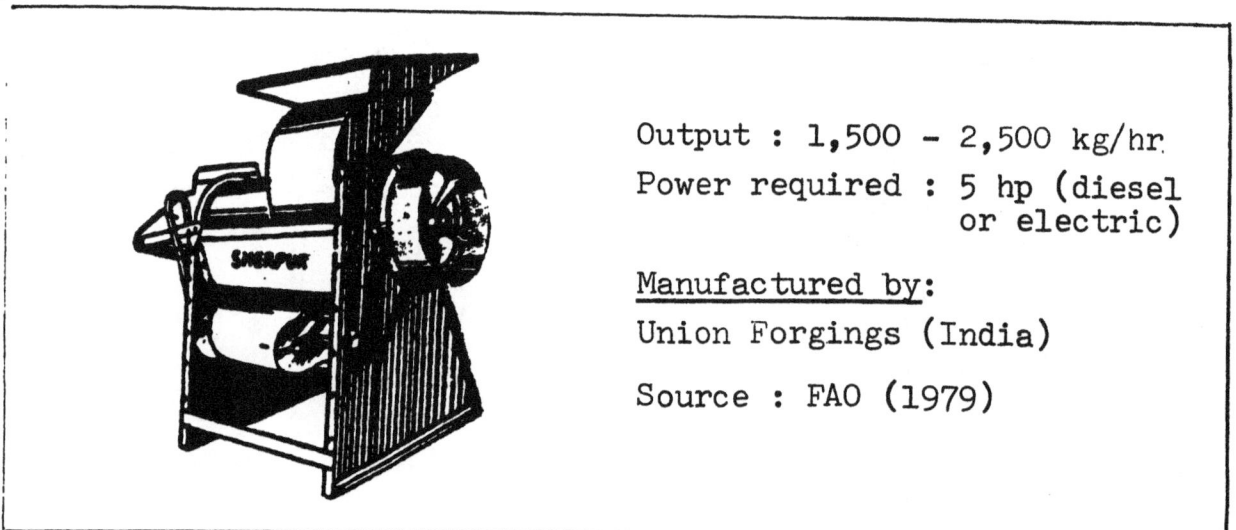

Output : 1,500 - 2,500 kg/hr

Power required : 5 hp (diesel
                 or electric)

Manufactured by:

Union Forgings (India)

Source : FAO (1979)

Plate III.12

"Sherpur" maize sheller

- 63 -

This sheller may shell husked or non-husked cobs. The revolving drum of the star and bar pattern, is flywheel-assisted and operates in conjunction with a hinged, stationary concave. The latter is spring-mounted at its outlet and is adjustable for varying sizes of cobs.

The majority of shelled maize passes through the concave bars and over a dust-extracting sieve as it gravitates to the winnowing hopper beneath. Cobs and husk, delivered by the 390 mm wide drum, are thoroughly shaken by a spring-balanced reciprocating rack which is fitted with a perforated floor. Loose maize is separated from the refuse by the shaker rack and returned to the winnowing hopper.

The maize falls to the auger for conveying to the bagging elevator and is cleaned of chaff, flake and broken cob by fan blast. An elevator takes the cleaned and shelled maize to the bagging-off spouts by means of an enclosed endless chain and slat carrier. Full provision is made for quick and efficient sack holding and changing.

Power required : 5 bhp.
Speed :   500 rpm
Pulley :   18" in diam (46 cm)
Output: - Husking and shelling: up to 2,700 kg
                   per hour;
                 - Shelling only: up to 2,600 kg per hour
Weight :   1,619 kg
Other crops: - Kaffir corn and bean attachments can
                   be provided at an extra charge.

Manufactured by: RANSOMES, SIMS AND JEFFERIES LTD.
                 Ipswich IP3 9QG
                 (United Kingdom)

Source : ITDG (1976)

Plate III.13

Large powered sheller ('Puma' sheller)

CHAPTER IV

MILLING TECHNIQUES

There is a wide range of milling techniques used at various scales of
production, and producing various types and qualities of maize meal. The
smallest scales of production are associated with techniques used at the
household level (e.g. use of mortar and pestle, of querns) while the largest
scales of production are generally achieved by roller mills. Intermediate
scales of production are associated with techniques used in stone, plate and
hammer mills. Given the focus of this memorandum on small-scale production,
this chapter emphasises these latter techniques, and only briefly reviews the
technology used by roller mills. Milling techniques used at the household
level are not considered, especially since an increasing number of rural women
tend to use the services of custom or community mills in order to be able to
devote more time to more productive activities.

I. PRE-TREATMENT OF MAIZE FOR MILLING

In general, untreated, shelled and dried maize grain is simply ground into
a meal or flour for the preparation of traditional products. However, some
traditional maize products require that the shelled grain be subjected to
various pre-treatment processes prior to grinding.

In Central and South America, maize is often used at the domestic or
commercial level for the preparation of tortilla, an unleavened bread. Prior
to grinding, the grain is boiled in a dilute solution of sodium and calcium
hydroxide for 15 to 20 minutes. This treatment loosens the bran and the tip
cap which are easily removed, together with any excess alkali, by washing. At
this stage, the slightly softened grain may be wet-milled in a mechanically
powered plate mill to produce the tortilla dough, "masa". Where the
requirement is for fresh tortillas, the "masa" is shaped and baked on the same

premises. Alternatively, after alkali treatment and washing, the slightly softened grain may be dried in oil-fired or electric ovens, ground in a hammer mill, packaged and marketed as instant tortilla flour. This pre-treatment gives the final product a characteristic and much desired flavour, and improves the nutritional properties of the grain.

Maize grain may also be lightly toasted prior to milling. In Central America, a particularly popular local drink or gruel - Pinol or Pinolillo - is a commercially prepared mixture of toasted maize whole meal (milled in a hammer mill), toasted cocoa beans, other cereals and flavourings such as nutmeg or cinnamon. The toasting process improves the flavour of the maize and increases its digestible energy by gelatinising the starch. It also retards the development of rancidity by inactivating the enzymes in the meal. Thus, the shelf-life of the product is further extended.

II. REVIEW OF MAIZE MILLING TECHNOLOGIES

There are two main milling technologies: one in which the grain is directly ground without any pre-processing and one in which the grain undergoes a number of pre-processing stages prior to milling. The former milling technology yields whole meal which contains both the bran and germ, while the latter one yields a large range of products including partly or fully de-germed meals called respectively bolted and super-sifted meals.

The production of whole meal is carried out in three types of mills: plate, stone and hammer mills. The output of these mills ranges from 25 kg per hour for plate mills to over 10,000 kg per hour for some large-scale hammer mills. The technical specifications of these mills are given in table IV.1. Plate, stone and hammer mills may use various sources of energy, including water-power, diesel and electricity. Some plate mills may use animal or wind power at relatively low outputs. The whole meal produced by these mills may be further sieved for the removal of large pieces of bran and germ. The mills may be equipped with grain cleaning equipment and attached to sieving devices. Water-powered mills are mostly custom mills while the other mechanically powered mills may be either custom or merchant mills, depending on the location and scale of production. The use of plate, stone or hammer mills is usually governed by local preferences, the intended scale of production and the type of output. Plate mills are extensively used in parts of West Africa (e.g. Ghana, Nigeria, Cameroon, Sierra Leone) whilst hammer

Cameroon, Sierra Leone) whilst hammer mills are more common in East Africa
(e.g. Tanzania, Kenya, Malawi). Stone mills for the dry grinding of maize
prevail in Central and South America, the Indian subcontinent, North Africa
and the Middle-East. Hammer mills are predominantly used for the production
of ground animal feed, such as in West Africa, Indonesia and Central America.

Table IV.1

Summary of technical data on mechanically
powered, plate, hammer and stone mills

Mill type / Characteristics	Plate	Hammer	Stone
Speed of rotation (rpm)	600	Up to 3600	600-800
Electric motor capacity (kW)	0.4-4	2-150	0.4-15
Diameter of grinding plates or stones (cm)	25	-	20-56 (v)* 61-71 (h)*
Average output per kW/kg/hr	67	74	80 (v)* 87-107 (h)*
Average output per hour (kg)	27-270	148-11,100	32-1,200(v)* 35-1,600(h)*

*v : vertical millstones
*h : horizontal millstones.

The second maize milling technology is used in roller mills where the
maize undergoes a series of pre-processing stages which include cleaning,
tempering, de-germing, sifting, reduction, etc. These mills yield a number of
maize products for various food preparations. The composition of some of
these products and the milling yields are shown in table IV.2. The range of
products generally varies from one mill to another depending on market

requirements. The extraction rate of the dry milled products is approximately 80 per cent. In general, the majority of the output comprises prime quality grits or meals with a fat content of less than 1.0 per cent. A small proportion of the output is made up of lower quality fines or flour with a fat content of 1.0 to 2.0 per cent. The by-products (i.e. the germ and bran) make up the remaining 20 per cent of the grain input. The germ is generally further processed for the extraction of oil while the bran is used for animal feed.

Table IV.2

Typical yields and composition
of de-germed maize products

Degermed products*	Milling yield (Per cent)	Typical particle size range (mm)	Moisture	Fat	Crude fibre	Ash	Protein
			- - - - - -(Per cent) - - - - - - - - -				
Flaking grits	12	3.4 - 5.8	14	0.7	0.4	0.4	8.4
Coarse grits	15	1.4 - 2.0	13	0.7	0.5	0.4	8.4
Medium grits	23	0.65 - 1.4	13	0.8	0.5	0.5	8.0
Granulated meals	3	0.30 - 0.65	12	1.2	0.5	0.6	7.6
Maize flour	4	Less than 0.2	12	2.0	0.7	0.7	6.6

* From a system yielding a multiple range of products. Milling yields and composition of products are dependent upon the individual processor's requirements.

The scales of operation of modern roller mills vary from 48 tonnes to 300 tonnes of maize input per 24 hours (2,000 kg to 12,500 kg per hour). However, smaller plants capable of processing as little as 7 to 9 tonnes of maize per 24 hours (300 kg to 400 kg per hour) are available, but are of limited distribution. The Indian Central Food Technological Research Institute (see Appendix II) has recently designed and developed a small-scale turnkey roller mill for the production of a limited range of maize products, particularly in rural areas. Details of design drawings are available for a nominal price directly from the Institute.

The following sections of this chapter provide detailed technical information on stone, plate and hammer mills. Roller mills are briefly covered since they are outside the scope of this memorandum.

III. WATER-POWERED MILLS

Water-powered mills are basically stone mills powered by the flow of water. A separate section (section IV.3) decribes stone mills powered by diesel engines and electricity.

The use of two circular flat stone surfaces moving in a horizontal plane, one above the other , forms the basis of the typical water-powered stone mill. It is commonly used in he highlands of East Africa, the Himalayas and in the Andean region where an abundance of fast-flowing streams provides the necessary motive power. Figure IV.1 shows a cross-section of a water-powered stone mill.

Water falling at an angle of approximately 80° and at a rate of approximately 5 litres per minute causes the rotation of a wooden veined paddle beneath the millstones. This motion provides direct drive - via a vertical wooden shaft connected to the paddle - to the uppermost of a pair of horizontal millstones which rotates at a speed of approximately 120 rpm (James, 1982; Temple, 1974; Ndambuki, 1981). The grain is fed through a hopper with a 15 cm diameter hole at its base (see figure IV.2). It then reaches the gap between the two millstones which are seated on a plinth. As the rotating top stone moves against the stationary stone, the grain is carried and ground as it flows from the centre to the periphery of the stones. Grooves cut into the stones assist this passage of the grain. The depth of these grooves decreases gradually from the centre of the stones outwards, thus allowing the gradual reduction of the grain into small fragments. The ground material falls into a channel surrounding the millstones from where it is collected. The gap between the stones can be adjusted by means of a simple wedge mechanism which alters the pressure on the top stone. In this way, flour of various textures may be produced.

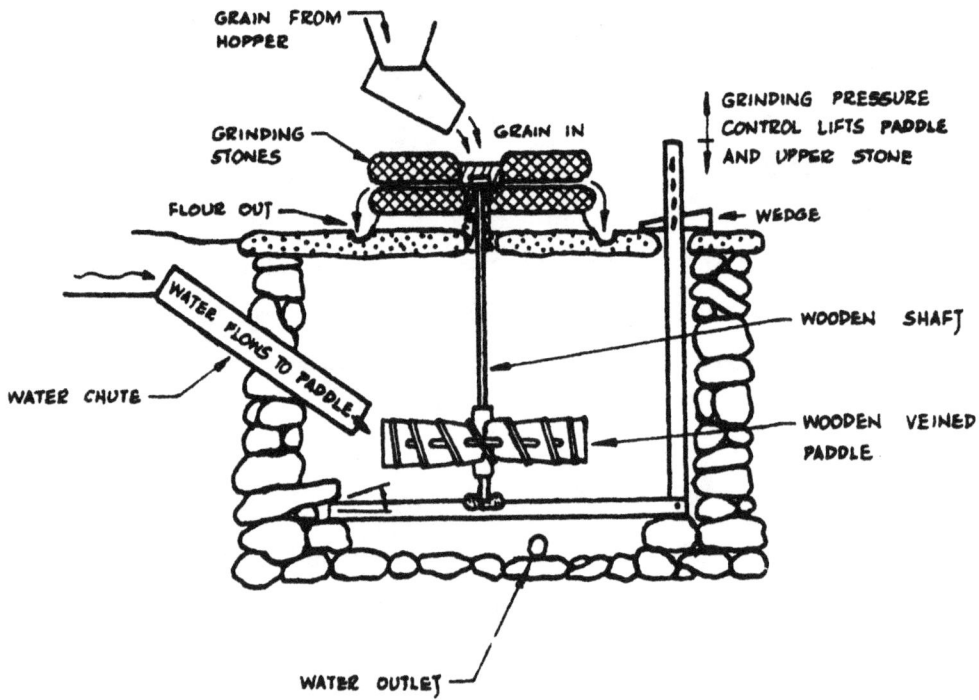

Figure IV.1

Cross-section of water-powered
stone mill

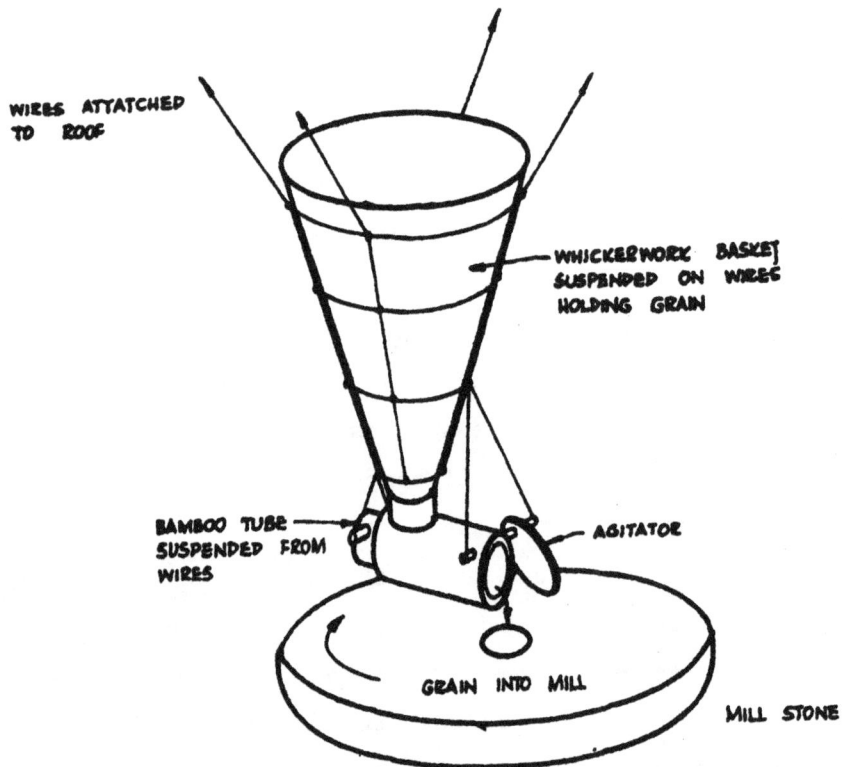

Figure IV.2

Grain feed hopper
Source : Temple (1974)

In a typical water-powered mill, with millstone diameters of approximately 75 cm, the output of the ground material ranges from 25 to 50 kg per hour depending upon the desired fineness of the material and the rotating speed of the stones. The latter, in turn, depends upon the rate of water flow.

Water-powered mills may be made from locally available materials by village craftsmen. The millstones themselves may be locally produced and dressed or, alternatively, imported.

IV. PLATE, HAMMER AND STONE MILLS

IV.1 Plate mills

Plate mills are made of a cast iron base to which are attached two enclosed vertical grinding plates (see figure IV.3). One plate is fixed while the other is belt-driven from an electric motor (0.4 to 4 kW), or diesel engine (in the range of 11 to 19 kW). The moving plate rotates at a speed of approximately 600 rpm. Some models may, alternatively, be driven from a tractor engine. The grain is screw-fed from a conical hopper into the gap between the two plates. This gap may be adjusted to vary the fineness of the ground material. The grinding plates, approximately 25 cm in diameter, are made from hardened cast steel. They are grooved to aid the shearing (cutting and crushing) and grinding of the grain. Different plates, with a range of groove sizes, may be used for the production of meals of varying textures. The hourly output of plate mills depends upon the required fineness of the product and the variety and moisture content of the original grain. Electric plate mills have an output of approximately 67 kg per kW per hour. Thus, a plate mill equipped with a 4 kW eletric motor may process approximately 270 kg of grain per hour. In parts of West Africa (e.g. Nigeria) and Central America, plate mills are used for the wet grinding of maize. For this purpose, plates with finer grooves than those used for dry milling are usually recommended by the manufacturer.

A few developing countries produce plate mills with imported engines, while other countries import the fully equipped mills. Plates IV.1 to IV.4 illustrate a few plate mills supplied by a number of manufacturers from both developing and developed countries.

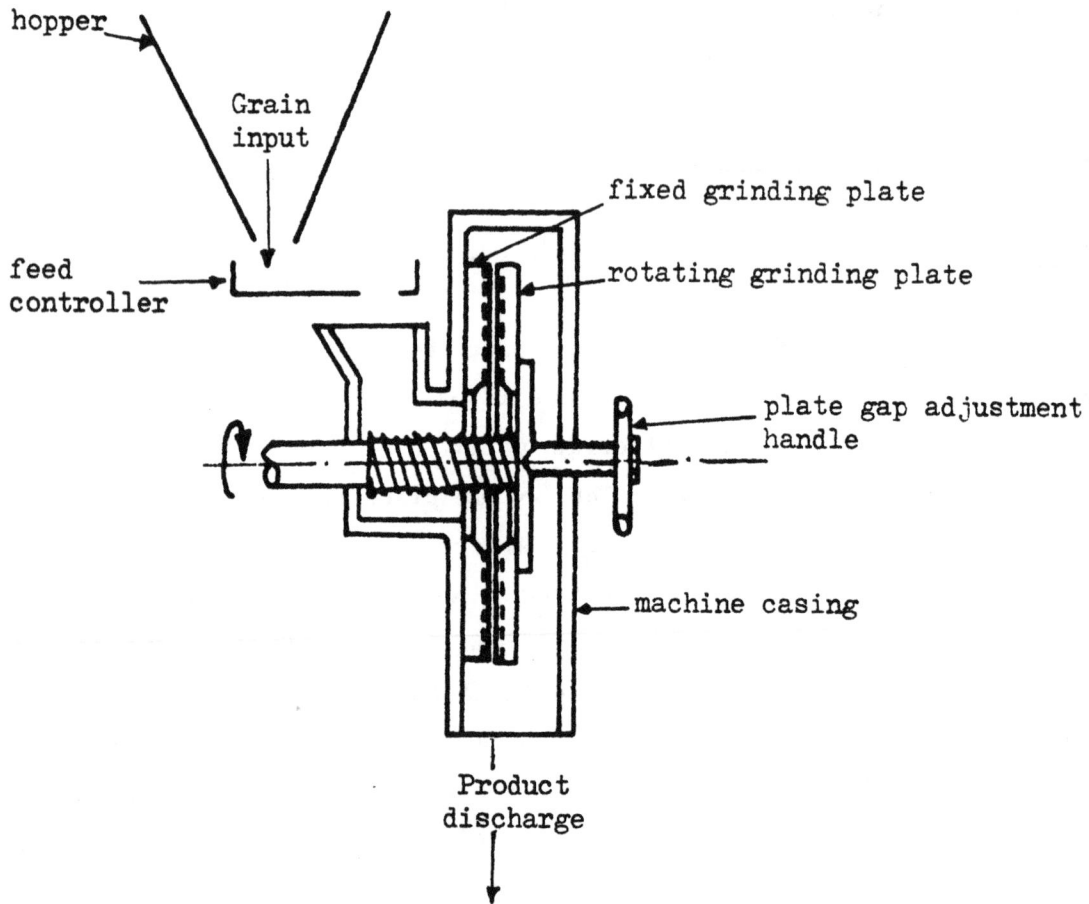

<u>Figure IV.3</u>

Diagrammatic representation of
a mechanical plate mill

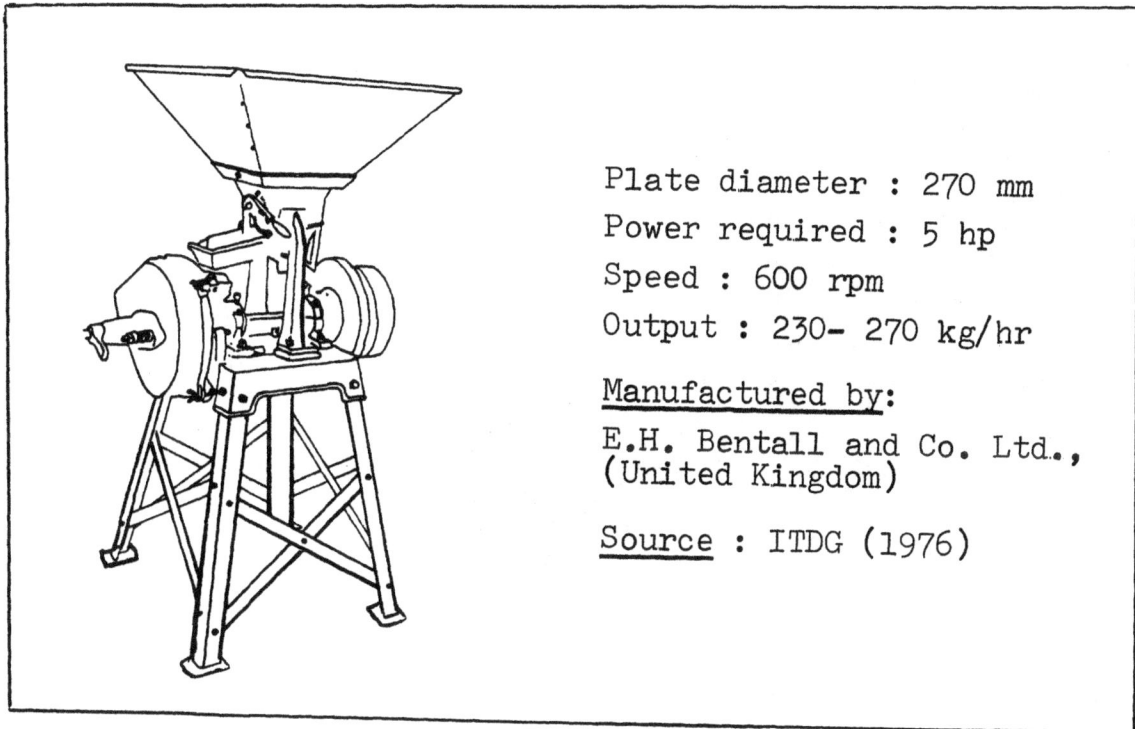

Plate diameter : 270 mm
Power required : 5 hp
Speed : 600 rpm
Output : 230- 270 kg/hr

Manufactured by:
E.H. Bentall and Co. Ltd.,
(United Kingdom)

Source : ITDG (1976)

Plate IV.1

"Superb" plate mill

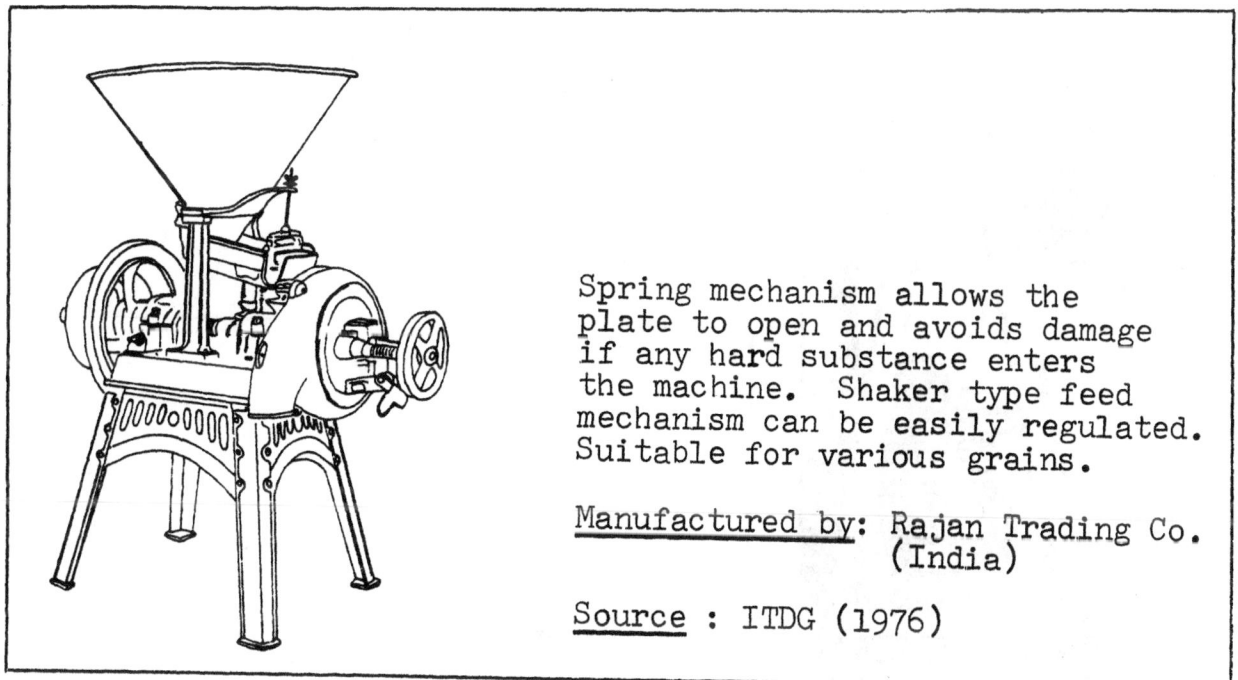

Spring mechanism allows the
plate to open and avoids damage
if any hard substance enters
the machine. Shaker type feed
mechanism can be easily regulated.
Suitable for various grains.

Manufactured by: Rajan Trading Co.
                 (India)

Source : ITDG (1976)

Plate IV.2

Amuda flat plate mill No. 1

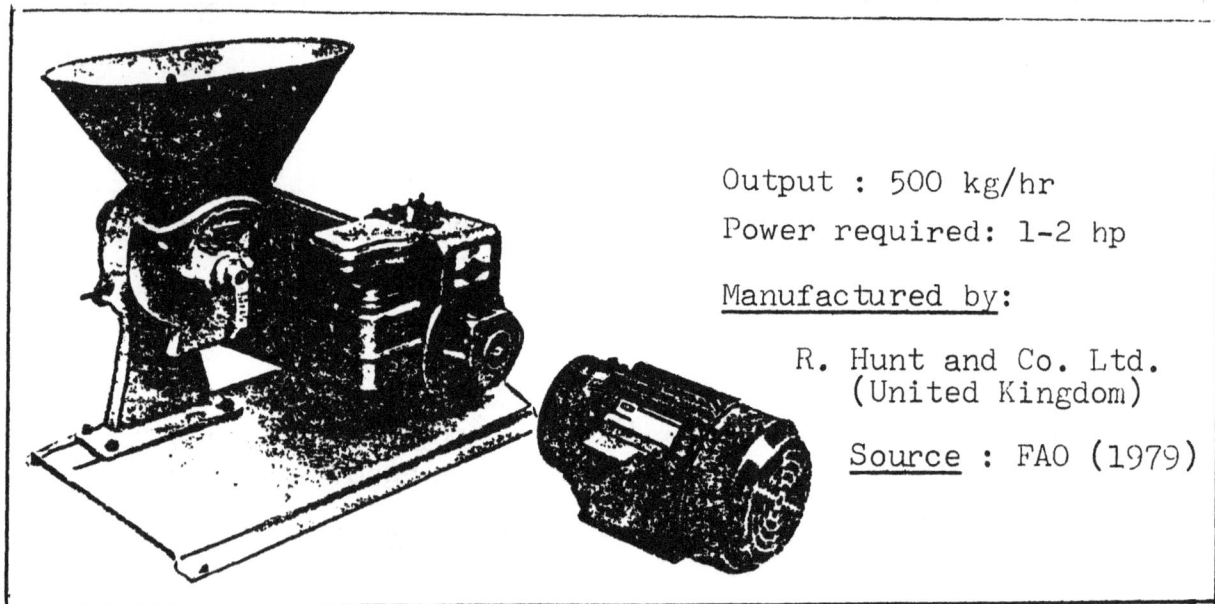

Output : 500 kg/hr

Power required: 1-2 hp

Manufactured by:

R. Hunt and Co. Ltd.
(United Kingdom)

Source : FAO (1979)

Plate IV.3

"Premier" 127 plate grinding mill

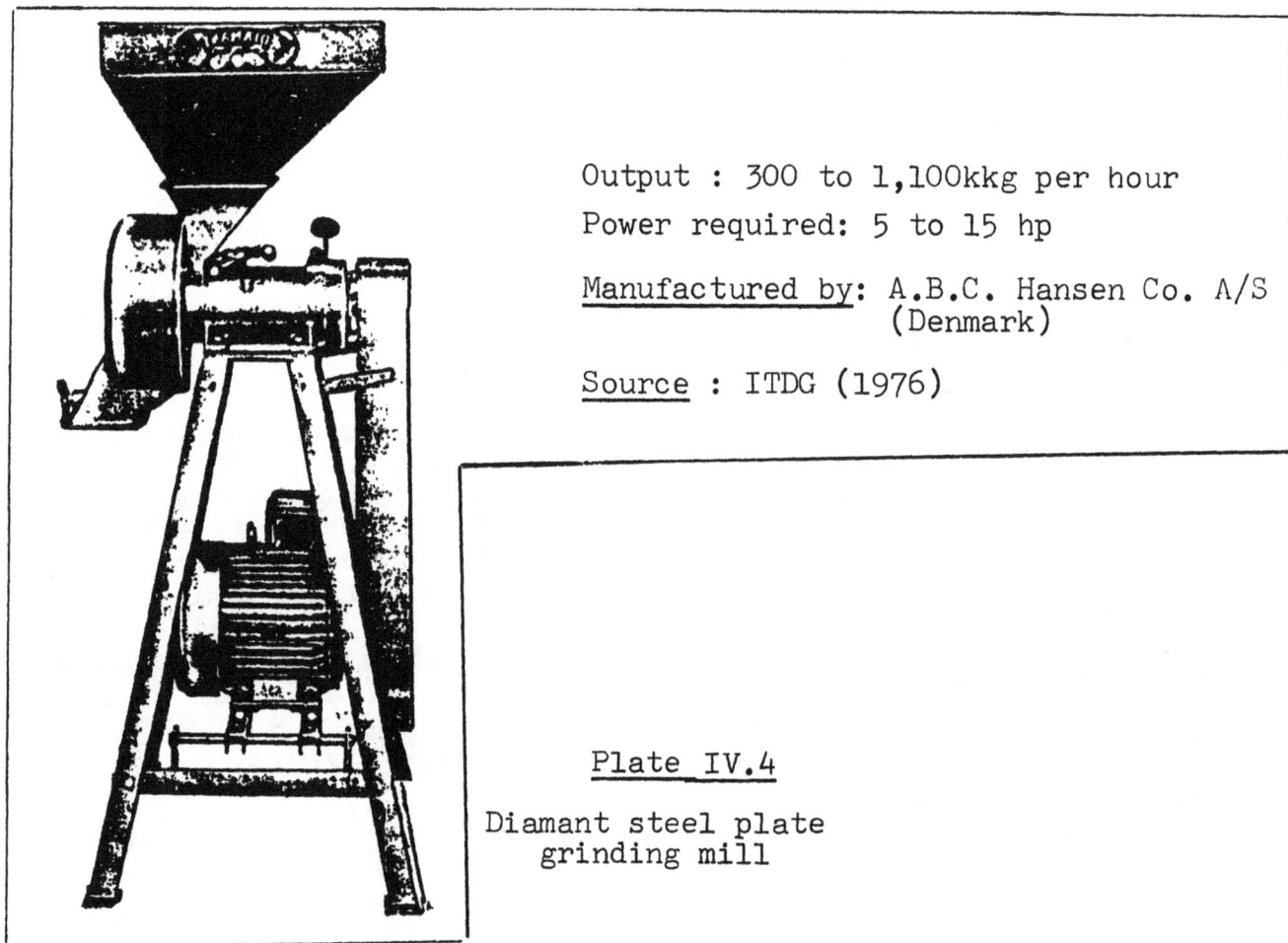

Output : 300 to 1,100kkg per hour

Power required: 5 to 15 hp

Manufactured by: A.B.C. Hansen Co. A/S
(Denmark)

Source : ITDG (1976)

Plate IV.4

Diamant steel plate
grinding mill

IV.2 Hammer mills

The hammer mills used in developing countries for maize dry milling are often imported from Europe or the United States. However, a growing number of these countries have started the manufacture of generally good quality hammer mills.

The design and capacity of hammer mills vary between manufacturers. In general, they comprise a cast iron body through which passes a horizontal rotary shaft powered by an external energy source (see figure IV.4). The latter is usually an electric motor or diesel engine. Occasionally, power is obtained from a tractor engine. The capacity of the electric motor varies from 2 to 150 kW depending upon the size and model of the mill.

A disc or discs, from which project short hammer-like plates, are attached to the end of the rotor shaft and enclosed in a metal casing. The hammer rotates at speeds of up to 3,600 rpm. They may be of the fixed or swinging type, and vary in number from 1 to 32. The fixed hammers are usually in the form of an iron casting whereas the swinging type are often made from heat-treated, 1.0 per cent chromium steel.

A screen, mounted on a fixed circular support, surrounds the hammers. The maize grain must be sufficiently reduced in size to pass through the screen before it is discharged from the milling chamber. A range of screens is available for the production of a variety of grades of ground material. A conical hopper, fixed above the milling chamber, holds the whole grain which is gravity fed into the mill.

Unlike the shearing action in the plate or stone mill, size reduction in a hammer mill occurs principally by impact as the grain hits the hammers, the metal of the screen, and the back wall and front casing of the mill. Impact also occurs between the grain itself. The grain is trapped and sheared between the hammer and the holes of the screen. The broken grain is retained in the milling chamber until its size is reduced sufficiently to allow its passage through the screen perforations.

The output of ground material varies according to the capacity of the motor, the size of the perforations in the screen and the variety and moisture content of the maize. As a general guide, the output per kW per hour is

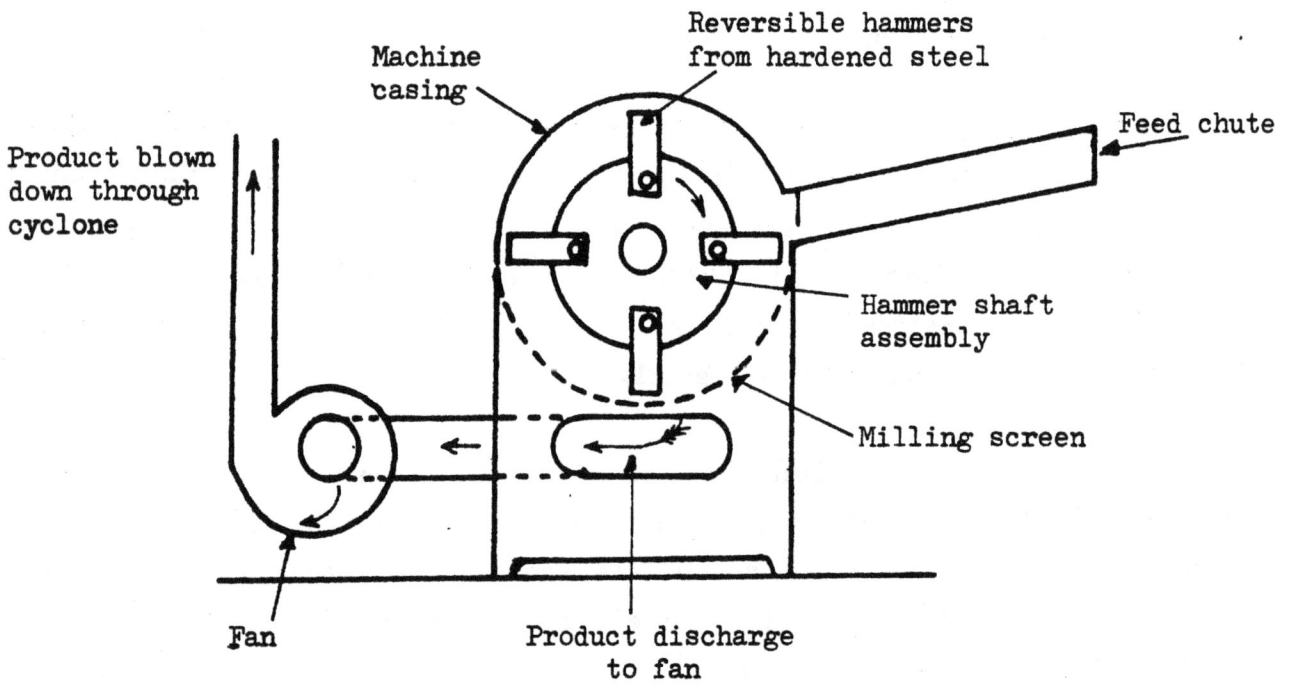

Machine
casing

Reversible hammers
from hardened steel

Feed chute

Product blown
down through
cyclone

Hammer shaft
assembly

Milling screen

Fan

Product discharge
to fan

Figure IV.4

Diagrammatic representation of
a hammer mill

approximately 74 kg for maize with a moisture content of 16 per cent and a screen with 3 mm holes. In the larger models (motor capacity greater than 5 kW), a cyclone discharges the ground material and cools both the mill and the product. In the smaller models (motor capacity less than 5 kW), the ground material is discharged by gravity from the base of the mill.

Plates IV.5 to IV.8 illustrate various hammer mills manufactured in both developed and developing countries. Table IV.3 indicates the physical characteristics of a selected number of hammer mills.

IV.3 Stone mills

In a typical stone mill, a conical or pyramid-shaped hopper holds the whole grain which enters the milling chamber through a feed valve. In some models, a shaking device and a screen prevent large impurities from entering the milling chamber. The milling of the grain is achieved by the shearing action of the flat surface of two millstones which are identical in size and construction. One stone is fixed to the milling chamber door while the other is mounted on a rotating drive shaft connected to an external energy source (e.g. an electric motor, diesel engine, or tractor engine). Figure IV.5 illustrates the basic design of a stone mill.

The grain from the hopper is fed, through the central hole in the rotating stone, into the gap between the two stones. As the rotating stone moves against the stationary stone, the grain is ground as it travels from the centre to the periphery of the stones. The two millstones may be set either horizontally with a vertical rotary shaft, or vertically with a horizontal rotary shaft. The vertical type is more common. It is shown in figure IV.5.

The diameter of the millstones varies according to model type and size. Generally, because of the weight of the stones and the relative difficulty in supporting them in an upright position, vertical millstones are smaller in diameter (20 to 56 cm) than horizontal millstones (61 to 71 cm). There are exceptions, however: some manufacturers produce vertical millstones of 71 cm and 81 cm diameter, while some horizontal millstones are only 30 cm and 41 cm in diameter. In the horizontal type, the crushed grain is moved to the periphery of the stones by centrifugal forces, whereas gravity assists the movement of crushed grain between the vertical millstones.

Table IV.3

Characteristics of selected hammer mills produced by manufacturers listed in Appendix I[1]

Manufacturer	Model	Type of "hammers"	Number of "hammers"	Use/main-tenance	Power (hp)	Rotation (rpm)	Suggested engine[2]	Output (kg/hr)	Cyclone attachment
ALVAN BLANCH	ESSEX MAJOR			Reversible	3-10			100- 300	
ETS CHAMPENOIS	REQUIN 4		24	Reversible*	7.5	3,000	E	500-1,000	
COMIA-FAO	BNT 4000				7.5-8-10	4,000	E, H	75- 150	
DDD PRESIDENT	MM		6	Re-useable**	3	3,000	E, 3	60- 100	
	MM/F		9	Re-useable**	2.5-3	3,000	E, 2.5		
	22		12	Re-useable**	5.5.	3,000	E, 5.5	200- 300	
	C2		24	Re-useable**	7.5	3,000	E	250- 500	
	B		30	Re-useable**	10	3,000	E	300- 800	
	MP				7.5,10		E	200- 500	
ELECTRA	BABY	Swinging	6	Reversible*	4.5;7; 7.5	6,000	E:4.5-7.5 D:14;P:7.5	150- 700	
	MINI	Swinging	6	Reversible*	2-3	3,000	P		
	VS1					6,000	H:14		Yes
GONRAD	T20	Swinging	20	Reversible*	4-8	5,000	E, H		Yes
	T24	Swinging	24	Reversible*	16-20	3,000	E		Yes
LAW	HBU4	Swinging	4		4;7.5		E	100- 600	
	EF				7.5	3,000	E	250	Yes
	Centaures	Swinging		Welding	12,15,25	3,000	E	500-2,000	Yes
	HPB	Swinging	4		6-15			100- 600	
	B15C				5.5	3,000	E:4kW	150- 250	
NDUMEE	ND20				12-25	4,000			
	ND30				16-100	3,600			
	GM40				25-100	2,000-2,600			
PROMILL	B2L			Interchan-geable	2,3,4, 5.5	1,500-3,000			
	B4C	Swinging	12		7.5,10, 15,20,25	3,000			
RENSON	BM12/55		12	Reversible*	5.5	3,000	E	100- 500	
	A5		15		5.5	2,800	E	300	
	B10		24		7.5		E	500	
	C15		36		10		E	700	
SECA ARGOUD	ALPIN	Swinging	6	Reversible*	4-5.5	6,000	P:7.5 D:14	150- 800	
	STOUT 27	Swinging	24	4 faces		3,000		400-1,500	
	EUROP 76			4 faces	5.5	3,000	E	600	Yes
	MIRACLE 71			4 faces	7.5	3,000	E, D	250- 400	
SKJOLD	SB	Swinging	16	Reversible*	4-10	3,800	E,D:11	300	Yes
	AM2	Swinging	12		10-13	3,800	E,D	120- 250	Yes
	BM2	Swinging	12	Reversible*	7.5-10	2,900		400	Yes
TIXIER	REIXIT BM	Swinging	15,18	Reversible	5.5-7.5	3,000	E	150- 700	
TOY	BA	Swinging			5.5;7.5 10	3,000		150- 500	
	T1		12	Reversible*	5.5;7.5; 8-10		E, P, D		Yes
SACM	PM 73				6	6,000	E,H	100- 200	
	BU 69				10	4,500	E,H	150- 300	
D.SECK		Fixed	6	Reversible	12	3,200	E, D		
SISMAR(SISCOMA)							E, P, D	250- 300	

1 Source : GRET (1983)
2 Letters in this colum designate the following engines : E for electric engines, D for diesel engines, H for heat engines and P for petrol engines. The numbers designate the capacity of the engine in hp.
* 4 faces
** 3 faces

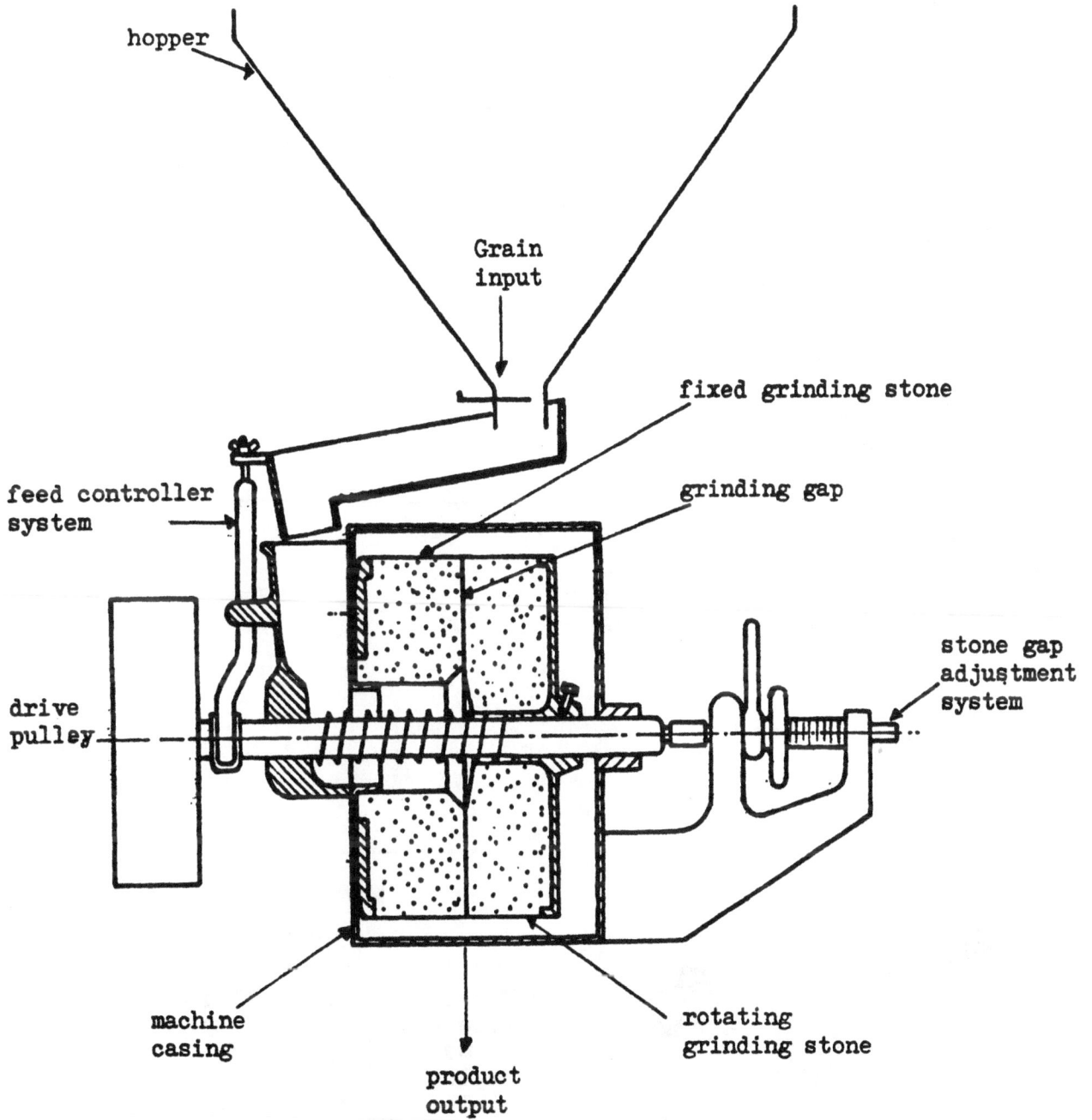

hopper

Grain
input

fixed grinding stone

feed controller
system

grinding gap

drive
pulley

stone gap
adjustment
system

machine
casing

rotating
grinding stone

product
output

Figure IV.5

Diagrammatic representation of a mechanical
stone mill with vertical grinding stones

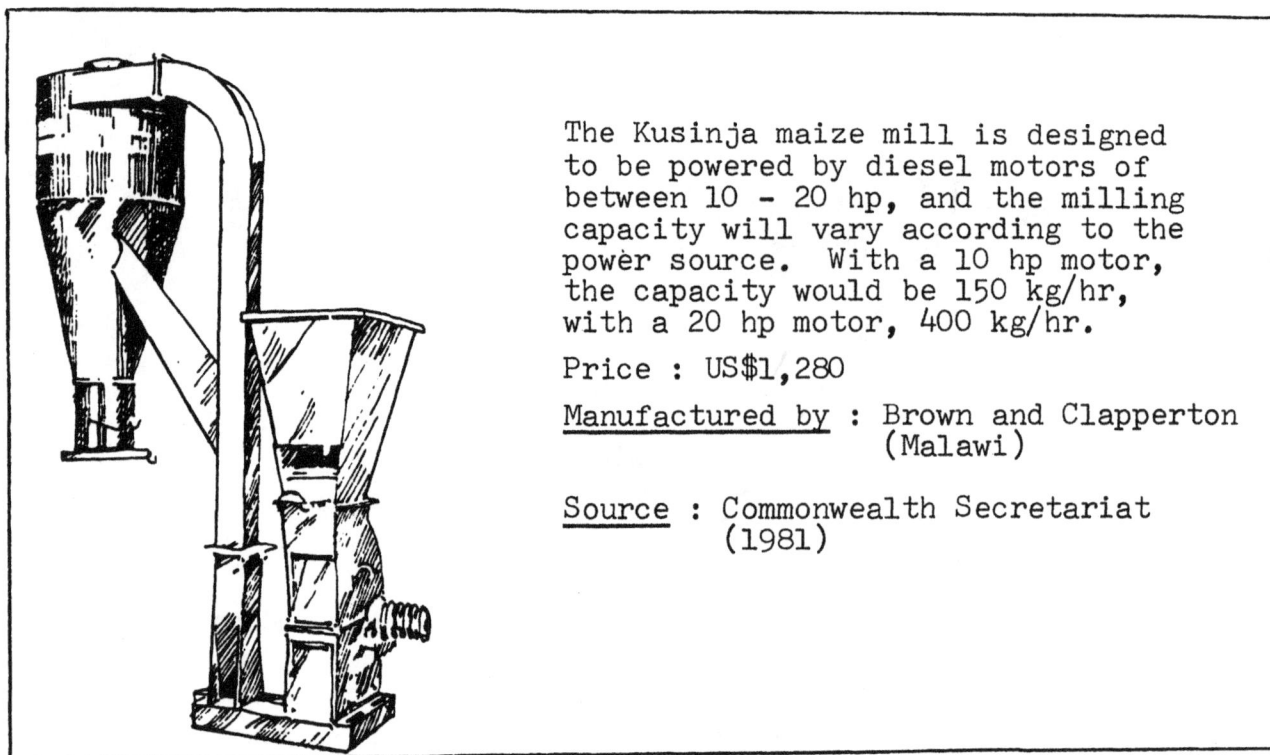

The Kusinja maize mill is designed
to be powered by diesel motors of
between 10 - 20 hp, and the milling
capacity will vary according to the
power source. With a 10 hp motor,
the capacity would be 150 kg/hr,
with a 20 hp motor, 400 kg/hr.

Price : US$1,280

Manufactured by : Brown and Clapperton
(Malawi)

Source : Commonwealth Secretariat
(1981)

Plate IV.5

Kusinja hammer mill

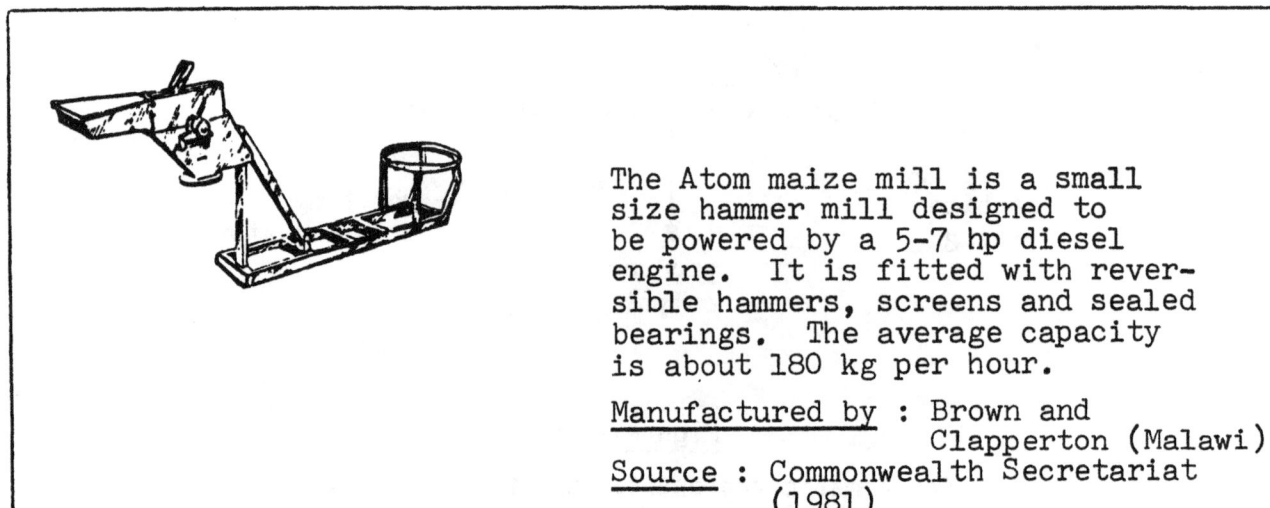

The Atom maize mill is a small
size hammer mill designed to
be powered by a 5-7 hp diesel
engine. It is fitted with reversible hammers, screens and sealed
bearings. The average capacity
is about 180 kg per hour.

Manufactured by : Brown and
Clapperton (Malawi)

Source : Commonwealth Secretariat
(1981)

Plate IV.6

Atom hammer mill

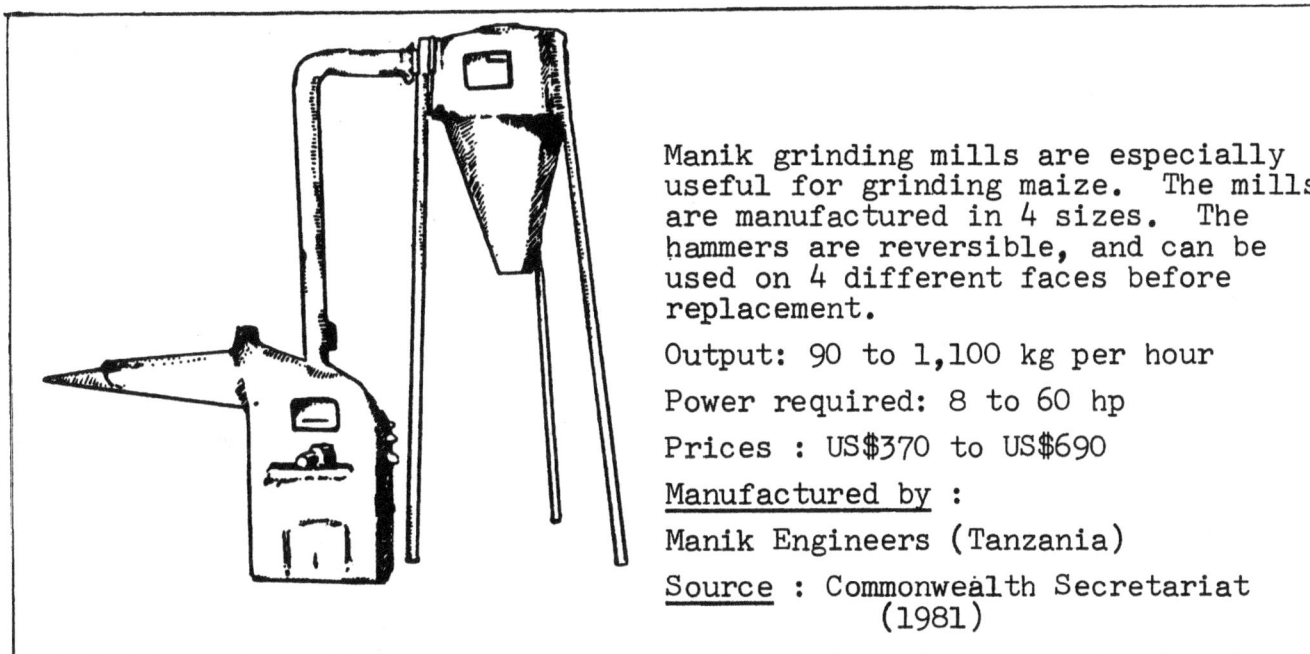

Manik grinding mills are especially
useful for grinding maize. The mills
are manufactured in 4 sizes. The
hammers are reversible, and can be
used on 4 different faces before
replacement.

Output: 90 to 1,100 kg per hour

Power required: 8 to 60 hp

Prices : US$370 to US$690

Manufactured by :

Manik Engineers (Tanzania)

Source : Commonwealth Secretariat
(1981)

Plate IV.7

"Manik" hammer mill

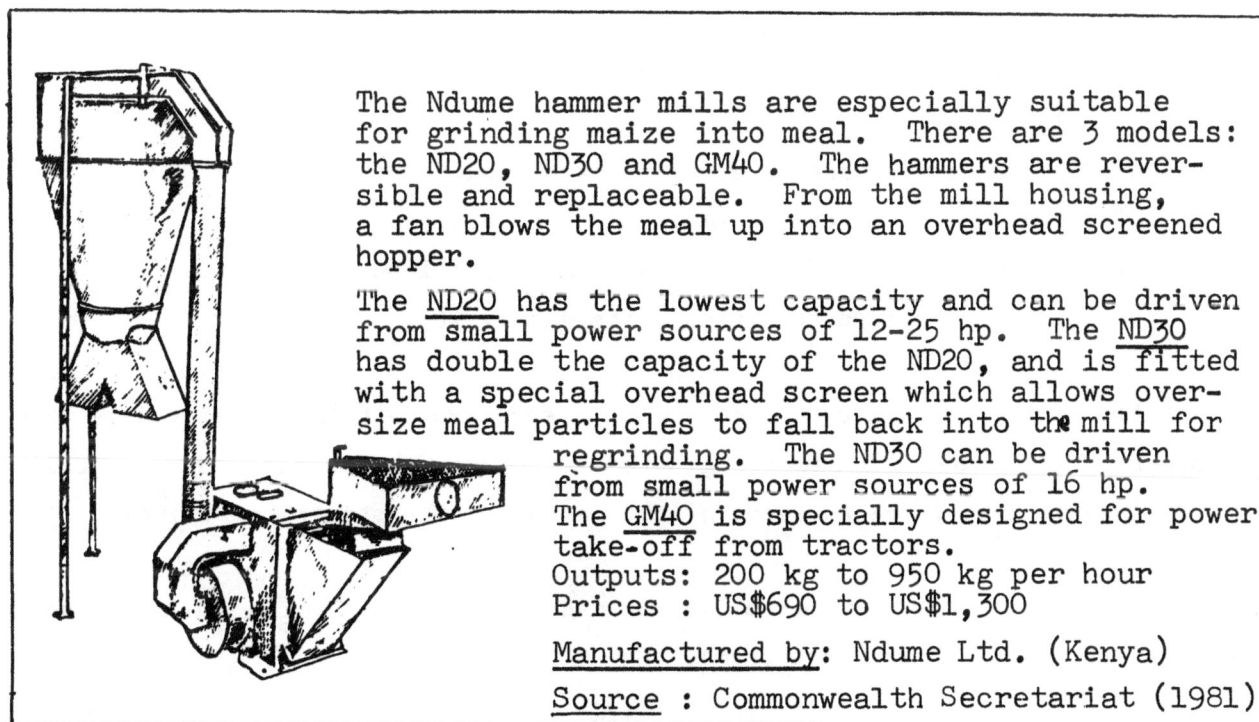

The Ndume hammer mills are especially suitable
for grinding maize into meal. There are 3 models:
the ND20, ND30 and GM40. The hammers are reversible and replaceable. From the mill housing,
a fan blows the meal up into an overhead screened
hopper.

The ND20 has the lowest capacity and can be driven
from small power sources of 12-25 hp. The ND30
has double the capacity of the ND20, and is fitted
with a special overhead screen which allows over-
size meal particles to fall back into the mill for
regrinding. The ND30 can be driven
from small power sources of 16 hp.
The GM40 is specially designed for power
take-off from tractors.
Outputs: 200 kg to 950 kg per hour
Prices : US$690 to US$1,300

Manufactured by: Ndume Ltd. (Kenya)

Source : Commonwealth Secretariat (1981)

Plate IV.8

Ndume power-driven hammer mill

The capacities of electric motors used in stone mills vary between 0.4 kW and 15 kW according to mill capacity and the diameter of the millstones. The motor capacity governs, in turn, the speed of rotation of the millstones within an optimum range of 600 to 800 rpm. The smaller diameter stones rotate faster than those of larger diameter. Thus, in a typical horizontal mill, the optimum rotation speed may be reduced to 400 or 500 rpm for stone diameters exceeding 61 cm.

The output of ground material depends upon the capacity of the motor, the speed of rotation, the diameter of the millstones, the variety of the grain and the desired fineness of the ground material. The average output of a vertical stone mill is 80 kg per kW per hour, while it may reach 107 kg per kW per hour in a horizontal mill equipped with large diameter stones. Thus, the average hourly output of stone mills varies between 33 kg and 1,600 kg per hour, depending on the motor capacity, the position (vertical or horizontal) and diameter of the millstones, the type of grain and the required fineness of the ground material.

Millstones are made out of one of the following materials:

- natural stones;
- small pieces of natural stones embedded in a matrix of cement
  or other suitable material. Other ingredients, such as emery,
  may also be added in the matrix; and
- artificial stones made of emery or carborundum, or a mixture
  of the above two materials embedded in a matrix of magnesium
  oxychloride cement. The carborundum may additionally be heat-
  treated or vitrified to increase its durability.

All types of millstones are usually enclosed within a supporting and protecting metal band. They are grooved to allow the shearing of the grain, as well as to assist the movement of the latter to the stones' periphery.

The casing of most stone mills is made out of cast iron although some models are made with a wooden frame.

A large number of developing countries manufacture stone mills for local use or for export to neighbouring countries. In many cases, the motor of these mills is imported.

Plates IV.9 to IV.14 illustrate various types of stone mills manufactured in developed and developing countries, while table IV.4 provides the characteristics of a selected number of mills.

IV.4 Efficiency of plate, hammer and stone mills

A comparison of the efficiency of the plate, hammer and stone mills shows that hammer mills are generally better suited than plate or stone mills for fine grinding. A plate mill generally consumes more power than a hammer mill during fine grinding, especially with grain at high initial moisture content. Plate mills would therefore seem to be more expensive to operate. A more efficient use of plate mills requires that the grain be pounded before milling: this is unnnecessary with the hammer mill.

Cyclones fitted to the large hammer mills cool the mill parts and the ground material. Their provision in plate mills or stone mills is unusual. However, as an increase in the temperature of ground maize may impair its nutritional characteristics and shelf-life, manufacturers of stone mills recommend an optimum rotation speed of the millstones which should not be exceeded by more than 25 per cent.

IV.5 Maintenance of plate, hammer and stone mills

All types of mechanical grinders require regular maintenance if they are to perform the grinding operation efficiently at all times. All moving parts require lubrication on a regular basis (e.g. weekly).

Most hammers, plates and millstones are reversible. Thus, they may be used for an extended period of time before sharpening, regrinding, dressing or replacement is necessary. Generally, the hammers need to be resharpened each week while the plates require regrinding every three to four weeks. Where excessive wear has taken place on some types of steel hammers, the tips can be returned to approximately their original dimensions by welding further metal. Using the correct materials, the new part may be made harder and thus more durable than the original. Natural stones wear more quickly than artificial stones and therefore need to be reversed or replaced more regularly. They are, however, cheaper to purchase. It should be emphasised that the life of

Table IV.4

Characteristics of selected stone mills produced by manufacturers listed in Appendix I[1]

Manufacturer	Model	Millstones		Use/main-tenance	Power (hp)	Rotation (rpm)	Suggested engine[2]	Output (kg/hr)[3]
		Material	Diameter (mm)					
ABC HANSEN CO.	DIAMANT	Artificial stone	250-550		1		E or D 6 to 30	
	FARMERS' FAVORITE	Artificial stone	600, 700		10	425	E or D .6 to 30	600
BENTALL	200 L090 SUPERB	Cast steel	267		5	600	D:11	250
ETS CHAMPENOIS	CLB	Cast steel	260	Reversible	4-6	850	E:4,H:4-6	60-180
	NOVA	Cast steel	160	Reversible	2.5-3	500-600	E:3,H:3	30
	DIAMANT H4	Corundum	500	Re-sharpen	3-4	550-600		120
	DIAMANT H6	Corundum	700	Re-sharpen	6-7			240
	V.300	Bakelite or metal	300 260	Replaceable	5.5-7.5	600-700	E	280-400
	V.400	Vit. cor. Metal	400 390	Replaceable	5.5-7.5	500-600	E, D, P	280-400
	JUNIOR	Hard cast steel	95	Reversible	0.5-7.5	100-125	E, H	25
COMIA FAO	BA 318	Vitrified corundum	300	Non-inter-changeable	4-6	750	E:4, H:5-6 D, P	80
	MB 317	Vitrified corundum		Non-inter-changeable	5-6	900	E:5.5,H:6	200
	ECLIPSE B30	Corundum	300		4-6	700-750		150-350
DANDEKAR MACHINE WORKS	DS style	Natural stone	160		6-8		E	250
DDD PRESIDENT	Nr.4/5/6/7	Natural stone				3-20	100-1,200	
	Nr.4/5 GM	Natural stone			5.5-7.5			
R. HUNT & CO	PREMIER 1A	Steel	254	Reversible	4	600	D: 7	150
	PREMIER 2A	Steel	305	Reversible	6	600	D: 11	200
IRUSWERKE	B/3/4/5/6		210-600		1.5-10		E	50-500
	RM/2/3/4/5/6		210-600		2-10		E	50-500
	RK/2/3/4/5/6		210-600		2-7.5		E	40-200
	CR2/3/4/5/6		210-600		2-10		E, D	40-250
MOULIS (CRICKET)	D4	Corundum	200-400	Non-revers.	2		E, P	100-400
RENSON & Cie	LE MODERNE	Corundum	300		4-5	400-600	E:7.5, P;D:7.5-11	200-300
	AVIMAT	Cast steel	90		.5		E: 5	60-120
	SILEX 113	Corundum			6-8		E: 4	350-600
	A320	Steel	320		4	750		600-1,000
SACM	MF 75				6-10	800-1,000	E, H	250
SAMAP	P220/380 (horiz.stones)	Stone	200	Interchang.	4	2,800	E	80-100
SECA ARGOUD	C300	Corundum	300		2-5	350	E: 3	250-400 l/h
	D400	Corundum	400		4-8	350	E: 5	150-200 l/h
	B205	Vit. corund.	200		3-4	350	E: 3/4	100 l/h
SICO GAUBERT	JUNIOR 170	Cor./emery	200		1-2	450	E	50-200
	SENIOR 170	Cor./emery	300		3	450	E	150-400
SIMON FRERES	N2GCV	Metal	218		1.5			85-200
	NGC 51	Metal	250		2-4			200-400
SKJOLD	KKE 16				6	650	D	300-400
TIXIER FRERES	REIXIT M9CV	Vit. corund.	250		3	700-800		150
	REIXIT M10CV	Vit. corund.	250		3	700-800		150-180
	REIXIT M11CV	Vit. corund.	300		5	700-800		200-300
	REIXIT M12CV	Vit. corund.	300		5	700-800		200-300
YAMAR		Corundum						400-500
		Corundum						600-700

[1] Source: GRET (1983)
[2] Letters in this column designate the following engines: E for electric engines, D for diesel engines, H for heat engines, and P for petrol engines. The numbers designate the capacity of the engines in hp.
[3] Output in k g/hr unless otherwise stated.

- 85 -

Output : 120 - 150 kg per hour
Power required: 3 - 4 hp
Manufactured by:
Etablissements Champenois (France)
Source : FAO (1979)

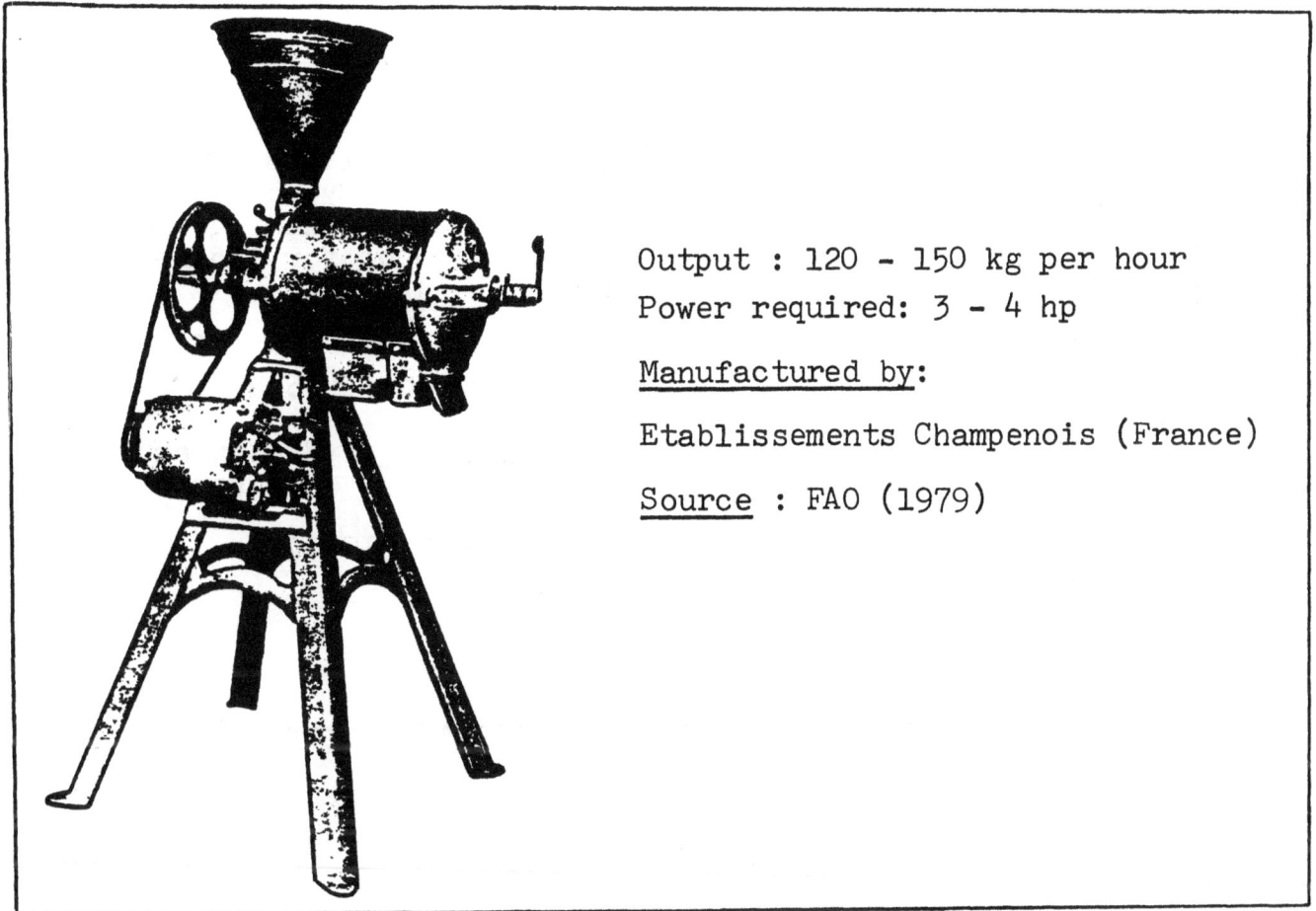

Plate IV.9
Horizontal stone mill

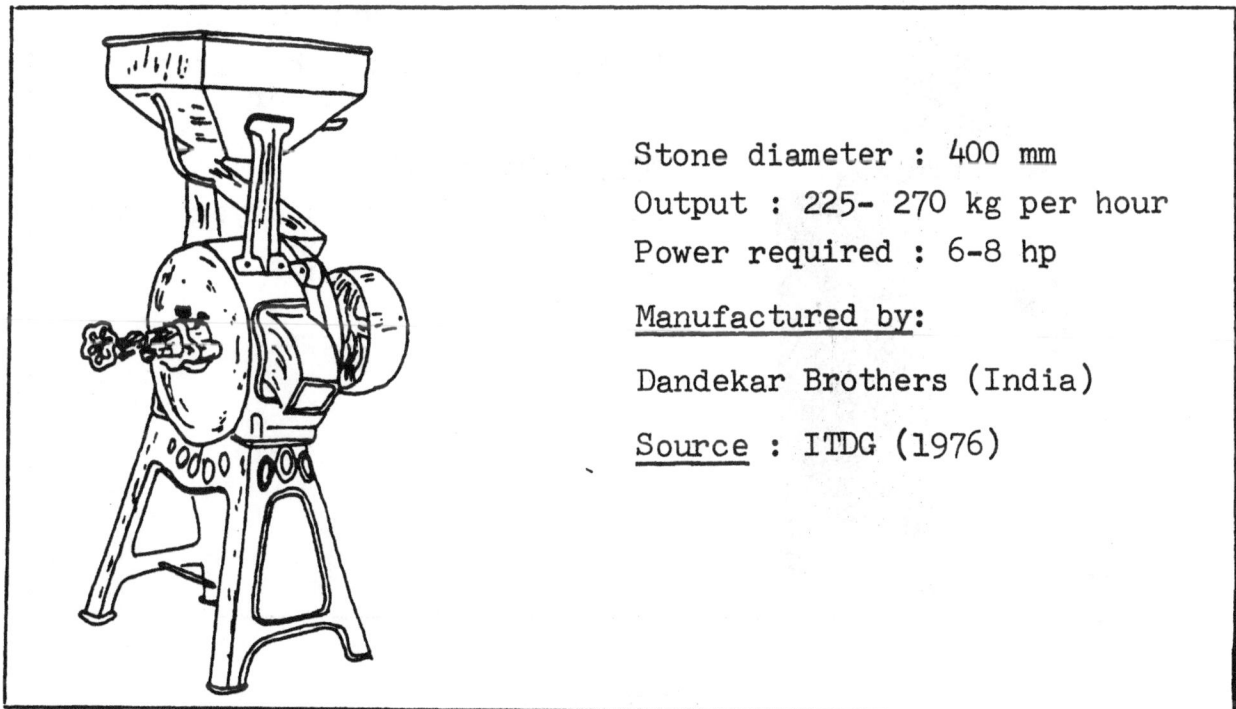

Stone diameter : 400 mm
Output : 225- 270 kg per hour
Power required : 6-8 hp
Manufactured by:
Dandekar Brothers (India)
Source : ITDG (1976)

Plate IV.10
Stone mill fitted with natural stones

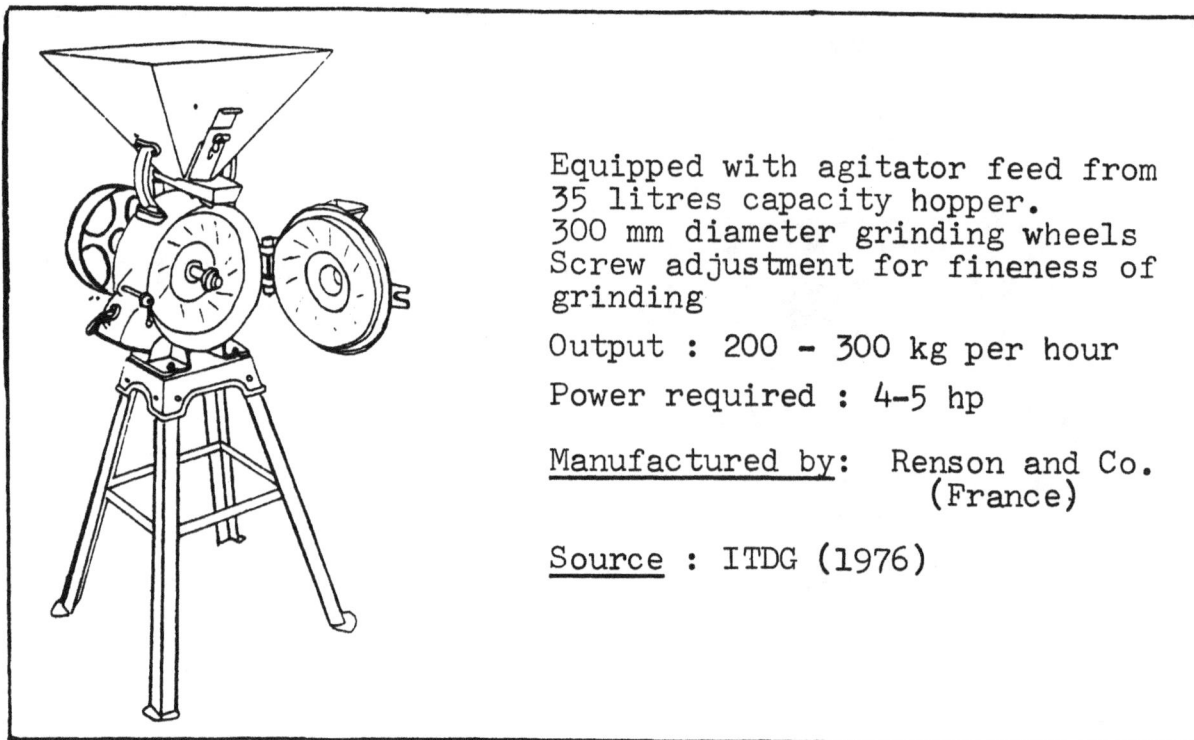

Equipped with agitator feed from
35 litres capacity hopper.
300 mm diameter grinding wheels
Screw adjustment for fineness of
grinding

Output : 200 - 300 kg per hour

Power required : 4-5 hp

Manufactured by: Renson and Co.
                        (France)

Source : ITDG (1976)

Plate IV.11

"Modern" stone mill

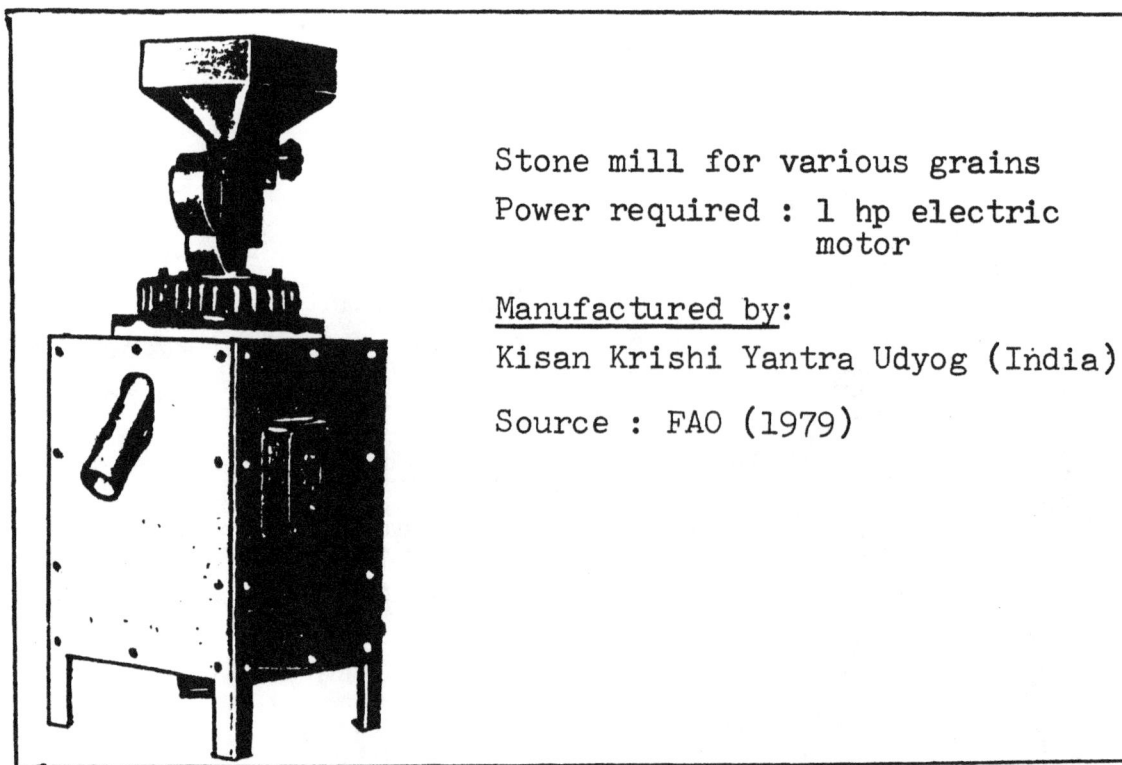

Stone mill for various grains

Power required : 1 hp electric
                         motor

Manufactured by:
Kisan Krishi Yantra Udyog (India)

Source : FAO (1979)

Plate IV.12

"Kisan" stone mill

- 87 -

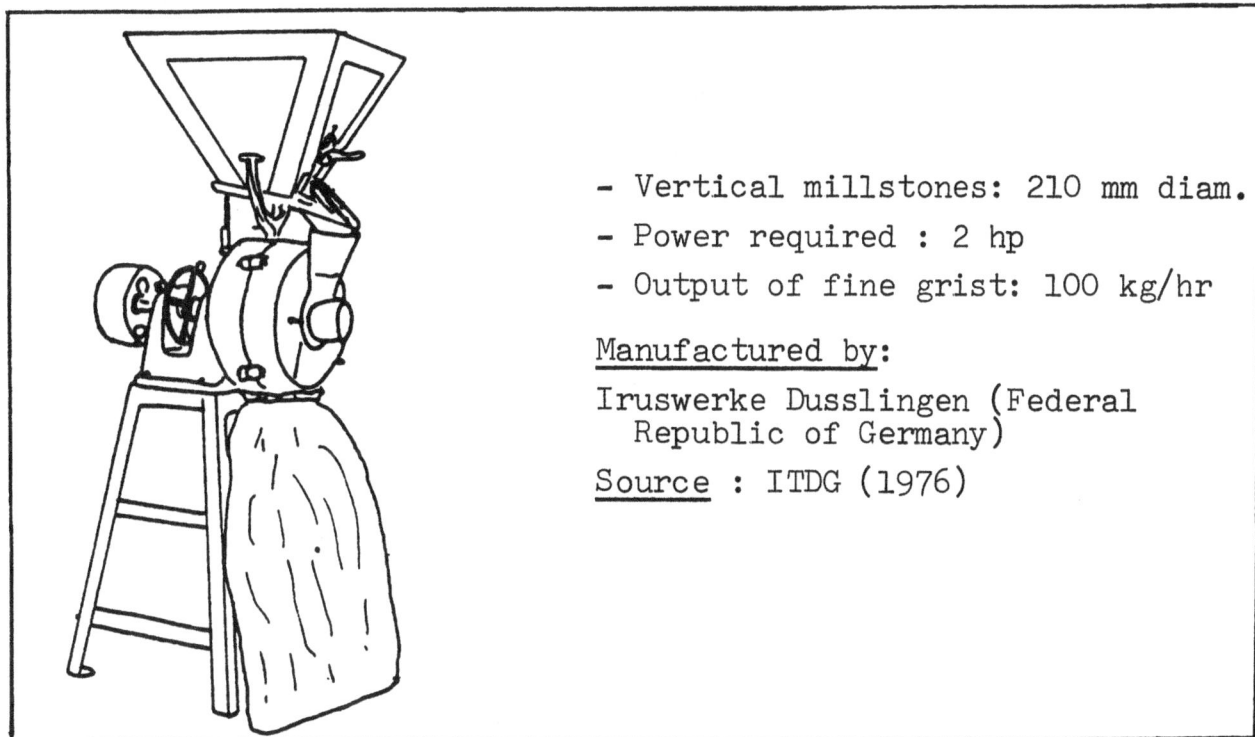

- Vertical millstones: 210 mm diam.
- Power required : 2 hp
- Output of fine grist: 100 kg/hr

Manufactured by:

Iruswerke Dusslingen (Federal Republic of Germany)

Source : ITDG (1976)

Plate IV.13

"R 2" grinding mill

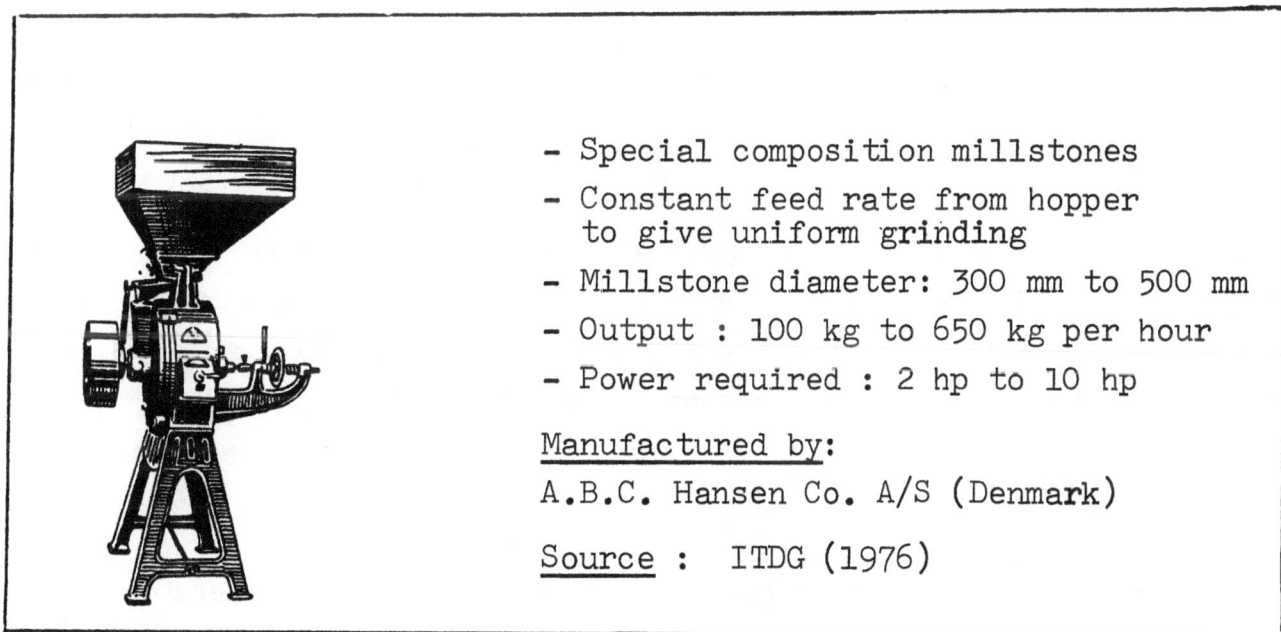

- Special composition millstones
- Constant feed rate from hopper to give uniform grinding
- Millstone diameter: 300 mm to 500 mm
- Output : 100 kg to 650 kg per hour
- Power required : 2 hp to 10 hp

Manufactured by:

A.B.C. Hansen Co. A/S (Denmark)

Source : ITDG (1976)

Plate IV.14

"Diamant" vertical stone mill

the milling parts, whether hammers, plates or stones, will be prolonged if foreign matter of mineral origin (e.g. fragments of stone or metal or sand) are removed from the grain prior to milling.

V. ROLLER MILLS

The main differences between roller mills and the other types of mills described in Sections III and IV are their larger output - exception made of small roller mills produced in India - and their capacity to produce a wide range of meal products. As already mentioned in Chapter I, it is argued that the shelf-life of these products is much longer than that of meal produced by traditional mills as a result of the removal of the germ.

Figure IV.6 shows the various operations undertaken in roller mills. These are:

- cleaning;
- tempering;
- de-germing;
- grading and aspirating;
- roller milling;
- sifting, purifying and aspirating; and
- packaging.

These operations are briefly described below.

Cleaning

The maize is cleaned to remove foreign matter of vegetable, animal and mineral origin by screening and aspiration. A magnet is included in the screening operation to remove metal fragments. A de-stoner (wet and dry) may be used, but is offered as an optional component of the mill.

Tempering

The grain must next be tempered before it enters the degerming system. The objects of tempering are to:

PRE-CLEANING
sufficient for animal
feed purposes

GRAIN INTAKE

↓

WEIGHING

↓

SCREENING & ASPIRATION ──→ LONG TERM STORAGE BINS

↓

DAILY STORAGE BINS

↓

WEIGHING

↓

SCREENING ASPIRATION

↓

DRY STONER - - - - - - - - -

↓

TEMPERING

↓

DAMP DE-GERMING   DRY DE-GERMING

↓←- - - - - - - - - - -

FINES ←──────── PLANSIFTERS

Animal feed

BRAN ←──────── ASPIRATORS

3 ENDOSPERM FRAGMENTS

CLEANED ENDOSPERM + GERM

CLEANING
additional process
for food uses

For oil
extraction ←───── GERM ←───── GRAVITY SEPARATORS

┌ - - - - - - - - - GRITS (ENDOSPERM FRAMENTS - - - - - ┐

PRIME QUALITY FLAKING
GRITS for conventional
manufacture of corn-
flakes and other pre-
cooked maize products

BREAK & REDUCTION GRADUAL
ROLLER SYSTEM

↓

PLANSIFTERS

LOWER QUALITY GRITS
for manufacture of
cornflakes and other
pre-cooked maize
products by extrusion

DRY DE-GERMED GRITS

PURIFIERS

RANGE OF DRY MILLED
MAIZE PRODUCTS ←- - - - - -

↓

PACKAGING

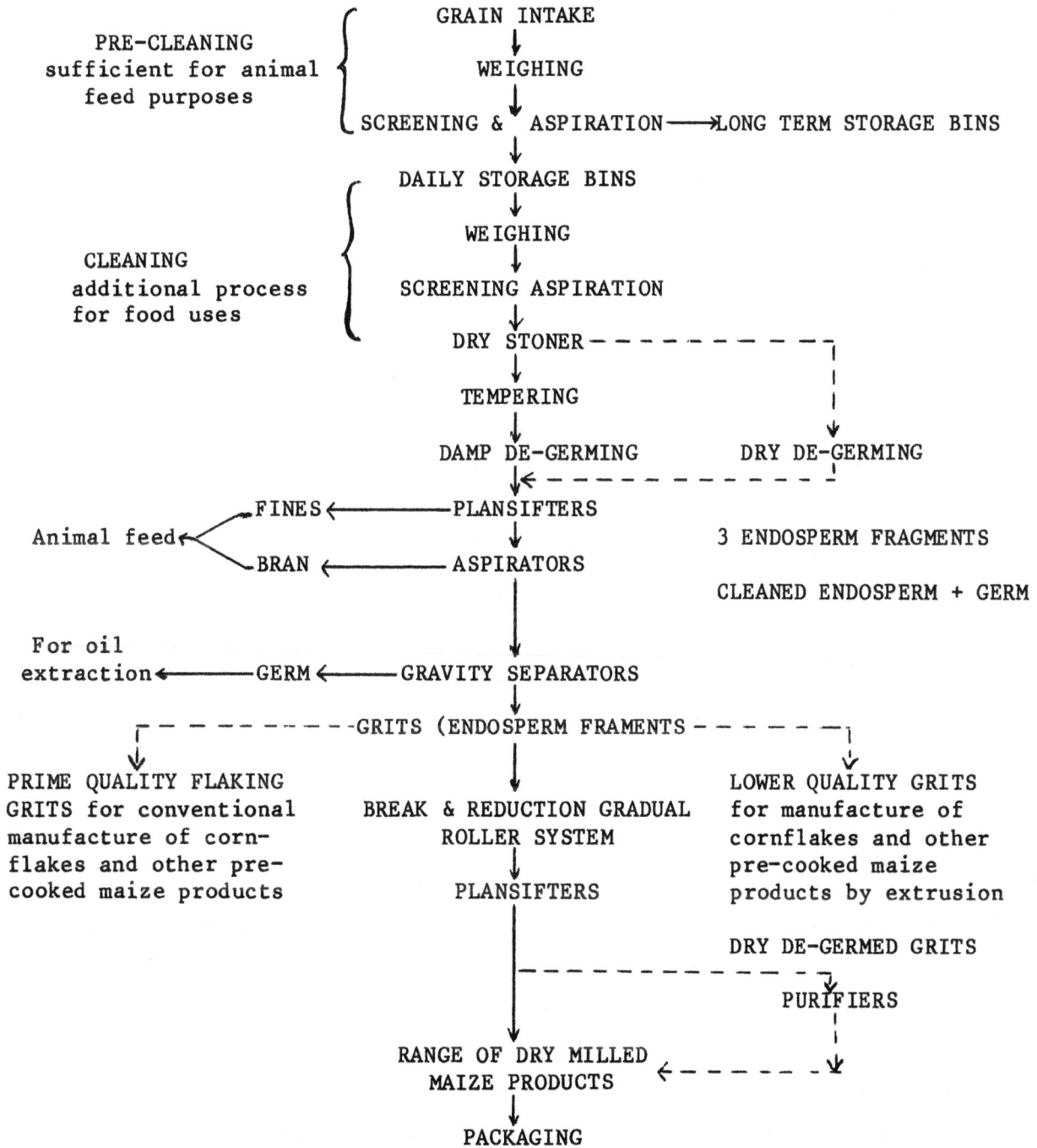

────────── Typical dry milling process.

- - - - - - Alternative process

Figure IV.6

Flow diagram of roller mill operations

- loosen the grain;
- moisten the germ with a view to making it more
  resilient and easier to separate;
- loosen and toughen the bran, making it easier to remove
  in large pieces; and
- moisten the endosperm for milling to allow for maximum
  production of grits and minimum production of flour.

Conventionally, tempering is achieved in a mixer conveyor by the addition of live steam or hot water to raise the moisture content of the outer layers of the maize grain from 14 to about 22 per cent. The moisture content of the endosperm increases only slightly since the grain is passed to the de-germer after little or no lying time. Alternatively, tempering may be carried out with cold water. This method is often used in developing countries where steam raising plants are expensive to install and operate. The maize is sometimes tempered for up to 16 hours to allow for moisture stabilisation at approximately 15 per cent. Subsequently, a second tempering of 10 to 20 minutes is used where 1 to 2 per cent of water is added to toughen the bran and germ. Alternatively, the maize is tempered for 4 to 5 hours to allow stabilisation at 16 per cent moisture content before the second tempering stage.

De-germing

De-germing, which includes decortication, is one of the key operations in roller milling. It is usually achieved by attrition. In the de-germer, the bran is scrubbed off, the germ loosened or excised and the kernel broken into two or more pieces as it passes between the moving and stationary parts of the de-germer. Two distinct streams of materials are produced: the larger particles, mainly broken endosperm and the finer particles of meal, bran and germ. Where the grain is initially tempered in steam or hot water, the fractions from the de-germer may require drying before further processing. With cold water tempering, moisture content are lower and the material from the de-germer may be processed immediately. As an alternative to the damp de-germing system, small plants (e.g. with a capacity of 2 tonnes per hour) may use a dry de-germination process. Whilst allowing the miller to produce a wide range of grits and meal economically, dry de-germing also removes the germ and the bran from the maize. Such a system offers a number of advantages, including a lower power consumption, less maintenance, and less tempering of the whole maize. Furthermore, it does not require a steam plant and   the drying of the finished products. It should be noted, however, that some tempering of the whole grain may be necessary to increase the moisture content to about 15 per cent. Otherwise, the endosperm would shatter during processing, thus giving too high a production of fine grits.

Other operations

Irrespective of the method of de-germination (damp or dry) the endosperm, bran and germ are subsequently sifted, aspirated and graded to remove the bran for animal feed formulations, the germ for oil extraction and the endosperm fragments for grinding.

Grinding of the endosperm into prime quality de-germed products is achieved on a long reduction roller system. Any remaining germ or bran adhering to the endosperm fragments are removed by aspiration throughout the reduction system. By sifting and purifying, the de-germed and ground endosperm fragments are separated into a variety of products according to local preferences. Depending upon the method of de-germing, the finished products may require drying to control their moisture content at a safe level for storage (usually 12 to 14 per cent).

Packaging

The greater flexibility between the use of labour and machinery in roller mills is found at the final, packaging stage. The degree of mechanisation that may usefully be applied to the packing system depends on the nature of the packaging material employed, the size of the package and the economic prices of factor inputs (Uhlig and Bhat, 1980).

Three alternative packing systems are available: manual, semi-automatic and automatic. These can be used equally well in small mills as in large mills. However, the capital cost of equipment and the staff required to operate it should be carefully considered in relation to the amount of material to be packed. Fully automatic packing lines operate with some surplus capacity even in the large mills.

The three packaging methods are briefly described below.

Manual packing

Manual packing includes the weighing of the product, the opening of the bag, the filling of the bag, the settling of the content and the shaping and sealing of the filled bags. One operative is required for each operation. The capital cost for this system is limited to simple seating accomodation for the packers, a working table area, trolleys for transferring the product from the outlet spouts to the packing area, and simple scales and hand scoops. The important features of a manual system are the low initial cost of equipment and the relatively high labour costs.

Automatic packing

The operations performed by a fully automatic packing line are the same as those for the manual system. All the operations and the transfer of material through this sequence is carried out by mechanised equipment without the need for intermediate handling by operatives. The direct labour requirements in this system are limited to one supervisor per shift.

Semi-automatic packing

In a semi-automatic packing line, the bags are presented to the automatic filler by hand instead of by mechanised equipment. The labour requirements are therefore increased to about two operatives per shift. In all other respects, this system is identical to the fully automatic packing line.

CHAPTER V

ORGANISATION OF PRODUCTION AND
INFRASTRUCTURAL REQUIREMENTS

The efficient organisation of production and the availability of an
adequate infrastructure constitute important requirements for the good running
of merchant mills which, in addition to milling the grain, need to store the
raw materials and the output, package the meal, etc. Furthermore, depending
on circumstances, small- to medium-scale merchant mills may also need to shell
and clean the maize grain, and undertake some limited sifting of the whole
meal for the removal of large pieces of bran and germ.

The above requirements are much more stringent in the case of large-scale
roller mills which require, in addition, an efficient management of the
plant. On the other hand, small custom mills, which only process shelled
maize brought by customers, require a simple infrastructure and little
management.

Since this memorandum is mostly concerned with the promotion of custom and
small-scale merchant mills, this chapter will focus on the organisation of
production, and the infrastructural and managerial requirements of the above
mills.

I. SKILL REQUIREMENTS

The running of milling equipment requires skills which may be acquired in
a relatively short time through, for example, on-the-job training. The mill
operators need only know how to adjust the mill for the production of various
types of meal and the processing of various types of grain. On the other
hand, the repair and maintenance of the equipment require mechanical skills
and, in some cases, a minimal knowledge of electricity. Skills are required
for the replacement of defective parts, the re-conditioning of hammers and
millstones, the maintenance of electric or petrol engines, etc. Furthermore,

improvements of the mill efficiency may require the design and manufacture of simple ancillary equipment.

The above maintenance, repairs or improvements require labour trained in technical schools or training centres. Additional on-the-job training in various mills may also be necessary.

The owners of custom mills should be responsible for all maintenance and repair activities since it would be too costly to hire skilled labour for this purpose. On the other hand, merchant mills may benefit from the full-time use of a skilled repair and maintenance operative, especially if the owner must handle managerial and supervisory tasks, as well as the buying of grains and the marketing of the output.

II. INFRASTRUCTURAL REQUIREMENTS

Infrastructural requirements depend on the type and capacity of the mill as well as on the need to store the raw materials and/or output. They also depend on whether the mill carries out pre-milling (e.g. shelling, cleaning) and/or post-milling (e.g. sifting, packaging) operations.

Depending on the adopted engine, mills may require an electric supply line or a petrol/diesel storage area. In areas where the supply of electricity is not dependable, mill owners should consider the use of petrol or diesel engines if they wish to avoid costly shutdowns of the mill. If electric power is much less expensive than other sources of energy, mill owners should assess the feasibility of investing in an electricity generator to compensate for electricity failures.

The buildings in which the milling equipment is installed and the storage areas should be well ventilated since the milling process generates a large amount of dust. The buildings should also be provided with a good lighting system in case the mill is operated on a two-shift basis.

Adequate roofing material should be used in order to avoid damage of the grain or whole meal by rainwater. The floor of the buildings should preferably be cemented to minimise the contamination of the whole meal by sand or dirt and to facilite the cleaning of the floor.

Space requirements

The space requirements for various types of mills and scales of production depend on the number of operations carried out by the mill (e.g. shelling, milling, sifting, bagging), the need for storing the raw material and/or output, the selected source of energy, etc. Thus, the space requirement for a given mill may not be estimated without detailed information on the characteristics and design of the mill. Consequently, this memorandum provides only indicative figures of space requirements rather than precise estimates of the latter for each type of mill.

The space requirement and a basic layout of six types of small-scale mills are provided in this section. The output of these mills varies from 1 to 8 tonnes per 8-hour working day. Table V.1 indicates the space requirements for each operation and storage area, for offices, etc. Pre- and post-milling operations, such as shelling, drying and sifting are not included. Additional space for these operations must be estimated if it is intended to carry out some or all of these operations.

Storage areas are calculated on the basis of 2 m^3 per tonne of grain in bags and 1 m^3 per tonne of whole meal. Grain stored in bulk requires less space, approximately 1.5 to 1.7 m^3 per tonne of grain. A requirement of six days' storage is assumed. Larger or smaller storage capacities may be required according to specific local conditions.

Figure V.1 illustrates the layout of an electricity-powered, horizontal stone mill with a capacity of 2 tonnes of maize per 8 hours (mill No. 2 in table V.1). It is intended to be a custom mill with no storage facilities since customers bring in grain and leave with whole meal.

Figure V.2 shows the layout of a similar mill which operates as a small-scale merchant mill as well as a custom mill (mill No. 3 in table V.1). In this case, the floor area is larger as both grain storage and product storage areas are needed.

Figure V.3 illustrates the layout of an electricity-powered hammer mill with a capacity of 4 tonnes of grain per 8-hour day (mill No. 5 in table V.1). This type of mill requires sufficiently large grain storage and product storage areas, as well as space for bagging/packaging and for an office. A mill with the same capacity, but powered by a diesel engine, requires additional space for the engine and a fuel tank (see figure V.4).

The layout of the other plants included in table V.1 are similar to those shown in figures V.1 to V.4.

Modern roller mills require a larger infrastructure. The factory layout is of paramount importance to ensure a maximum space utilisation. Figure V.5 illustrates the layout of a three-storey, specially designed roller mill with a capacity of 2 tonnes per hour (the material is gravity-fed from one operation to the next). However, a single storey building of larger floor area could be used if it is sufficiently high for the elevation of the materials through the various milling stages.

Table V.1

Suggested space requirements for various types of small-scale mills

		Floor space (m^2)					
Mill number		1	2	3	4	5	6
Type*		P	S(h)	S(h)	H	H	H
Capacity of mill (tonnes/8 hr)		1	2	2	4	4	8
1.	Mill with electric motor	1	1.5	1.5	4	4	9
2.	Grain feed in	1	1.5	1.5	2	2	4
3.	Product bagging/packing	1	1.0	2.0	2	4	6
4.	Grain storage	–	–	6.0	–	12	24
5.	Product storage	–	–	3.0	–	6	12
6.	Work space, including maintenance	2	2.5	2.5	4	7	7
7.	Spares	1	1.0	1.0	1	1	2
8.	Office	–	–	2.5	–	6	6
9.	Diesel engine room	–	–	–	10	10	12
10.	Fuel tanks	–	–	–	4	4	4
11.	Total with electric motor	6	7.5	20.0	13	42	70
12.	Total with diesel engine	–	–	–	27	56	86

* P : Plate mill

  S : Stone mill

  (h): Horizontal

  H : Hammer mill

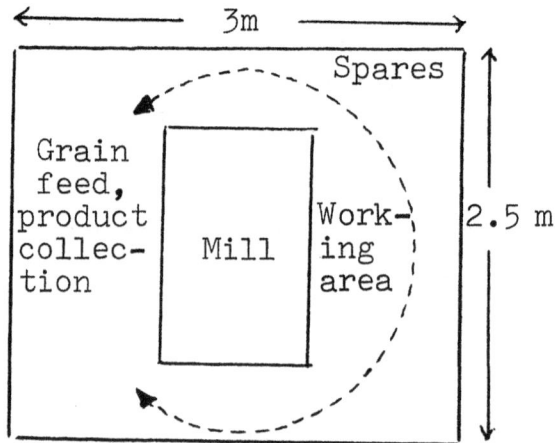

Scale: 1cm = 0.5m
Total area: 7.5m^2

Figure V.1

Custom-operated, electricity-powered horizontal stone mill -
2 tonnes maize per 8 hours

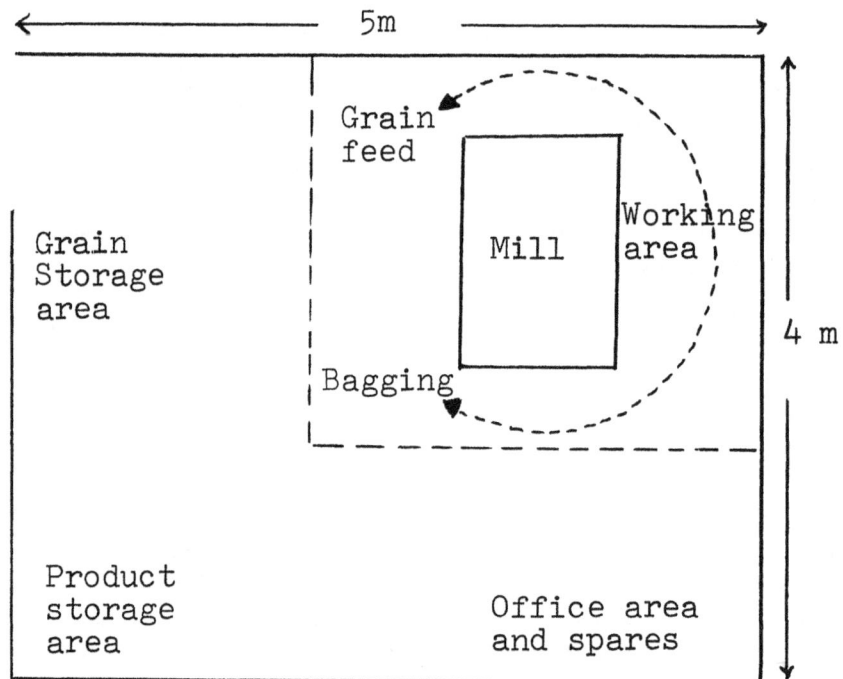

Scale: 1cm = 0.5m
Total area: 20m^2

Figure V.2

Commercially-operated, electricity-powered horizontal stone mill,
2 tonnes maize per 8 hours

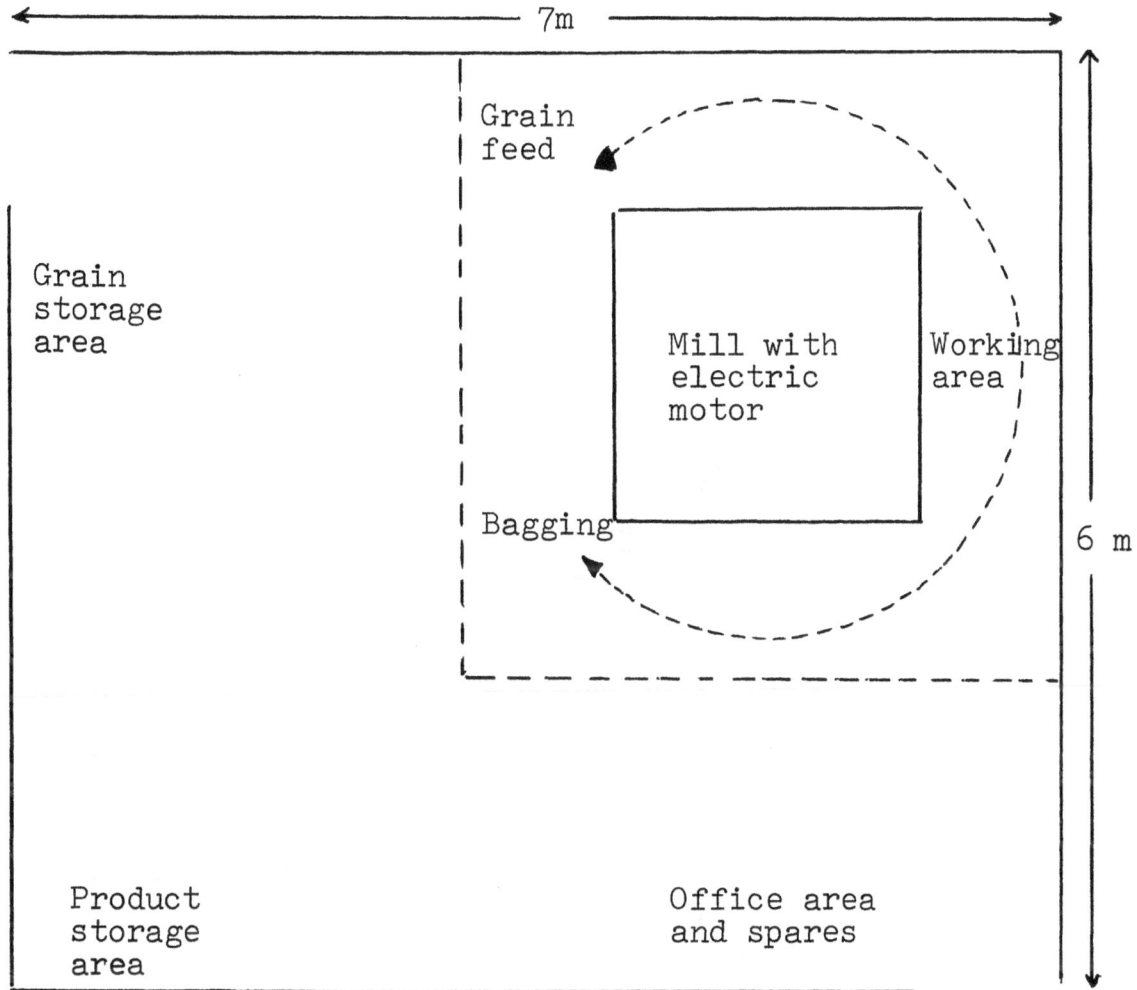

Scale: 1cm = 0.5m
Total area: 42m^2

Figure V.3

Commercially-operated, electricity-powered hammer mill -
4 tonnes maize per 8 hours

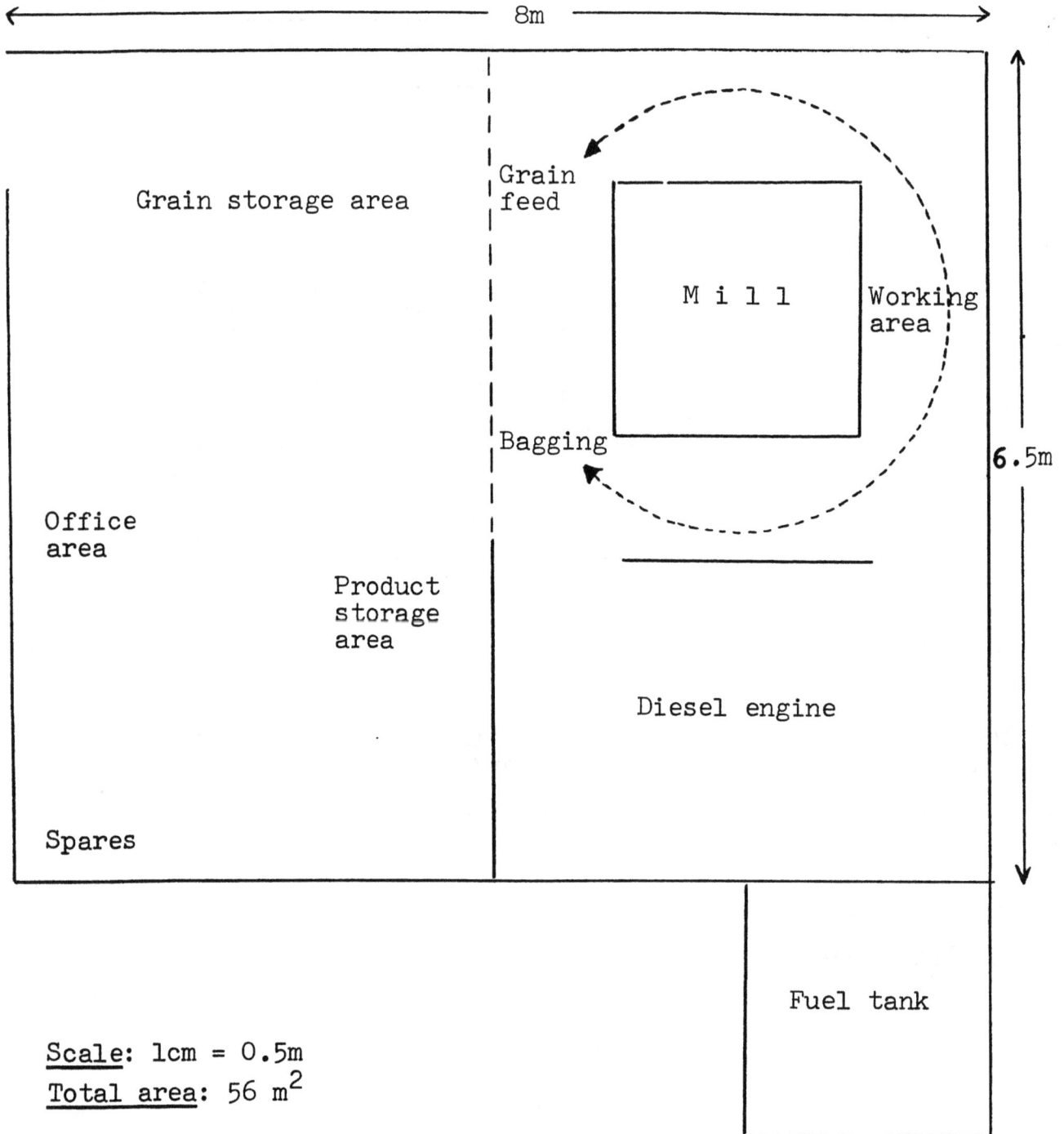

Figure V.4

Commercially-operated, diesel-powered hammer mill -
4 tonnes maize per 8 hours

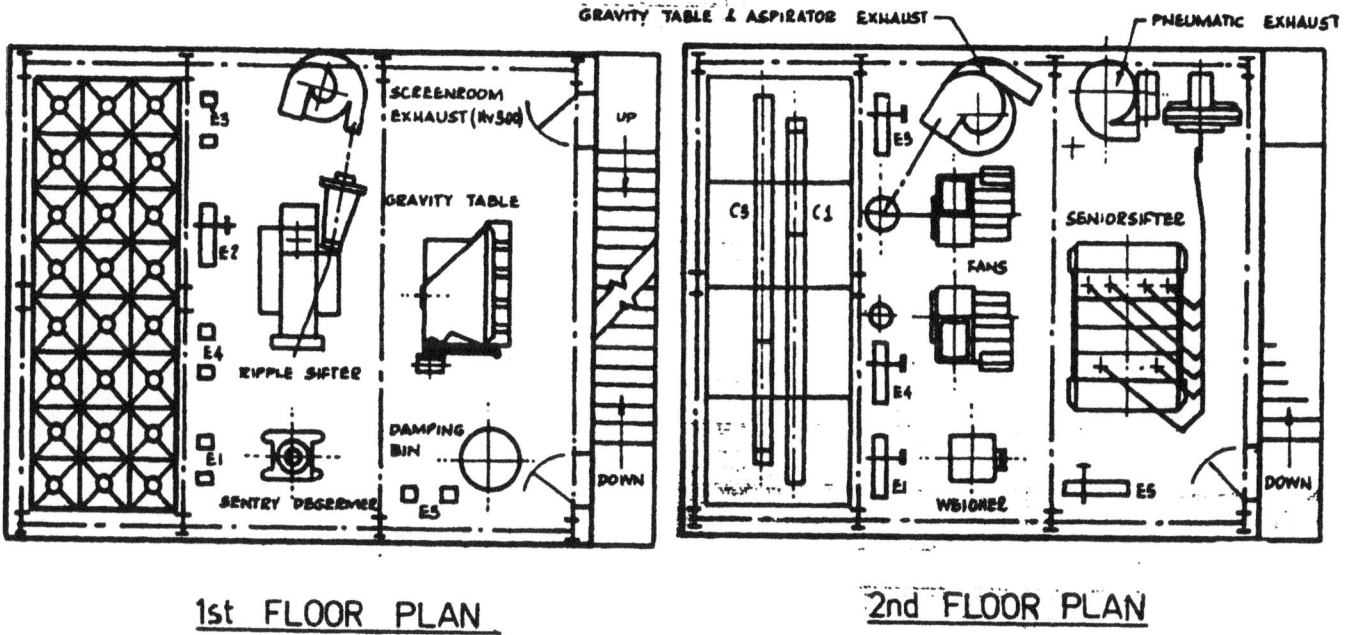

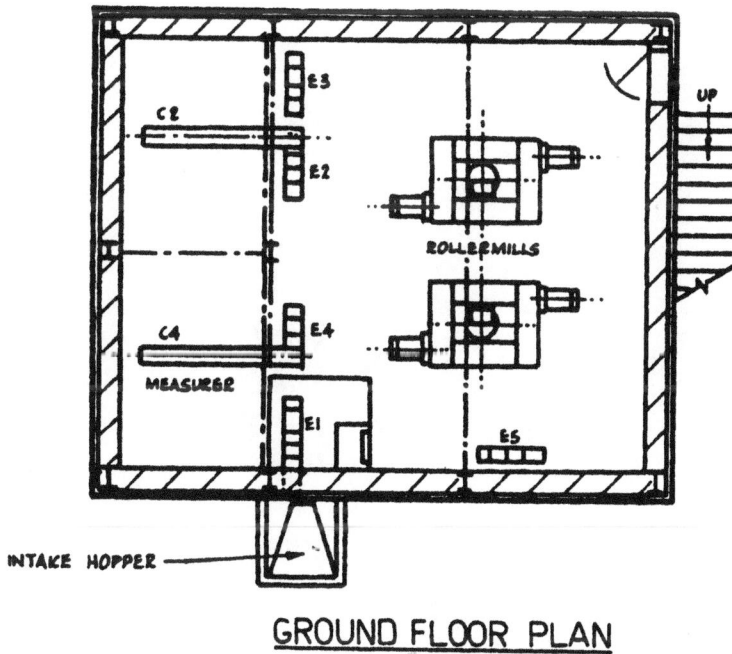

1st FLOOR PLAN

2nd FLOOR PLAN

GROUND FLOOR PLAN

Figure V.5

Floor plan of a three-storey roller mill -
2 tonnes maize per hour

CHAPTER VI

METHODOLOGICAL FRAMEWORK FOR THE
ESTIMATION OF PRODUCTION COSTS

I. INTRODUCTION

An entrepreneur wishing to invest in a maize milling plant does not generally have many choices regarding the range of scales of production. This range is, by necessity, determined by the investment funds available to the entrepreneur and market demand. Thus, these two factors determine whether he will invest in a custom mill, small-scale merchant mill, or medium to large-scale hammer or roller mill. However, within a given range of scales (e.g. small-scale merchant mills), the entrepreneur may choose among various types and capacities of mills (e.g. small-scale stone, plate or hammer mills). Furthermore, he may wish to investigate if it will be profitable to carry out various pre-milling (e.g. shelling) and/or post-milling (e.g. sifting, packaging) operations. Thus, given the adopted range of scales, the miller will need to estimate the costs and revenues associated with each alternative milling technology within this range to identify the most profitable alternative. The purpose of this chapter is to describe a relatively simple methodology for the economic evaluation of alternative maize milling technologies. This methodology applies mostly to small-scale custom and merchant mills. Entrepreneurs wishing to invest in a large-scale hammer or roller mill are advised to secure the services of a qualified engineering firm to obtain a reliable estimate of the profitability of the mill.

II. DETERMINATION OF THE SCALE OF
PRODUCTION AND TYPE OF OUTPUT

The determination of the scale of production requires, as a first step, the identification of the market to be supplied. For small-scale mills, with an output of 1 to 8 tonnes per day, the potential market area may include a village or group of villages, a district or a region of the country. The region is probably the largest potential market area for mills with a capacity of 8 tonnes per day. A national market is generally associated with large-scale mills.

Once the potential market area has been determined, the entrepreneur will need to estimate the volume of maize to be processed for this market. This estimation requires information on maize meal consumption in the market area. If this information is not available, the entrepreneur will need to undertake his own marketing study to avoid an under-utilisation of the mill's capacity in case market demand is overstated. This estimation of market demand should take into consideration other sources of supply of maize meal such as already established local mills or meal imported from outside the market area. The entrepreneur may thus estimate his potential share of the market.

Once potential demand and market share have been estimated, the would-be miller should investigate whether the required supply of maize grain will be available. This is an important consideration since many small-scale mills in developing countries have been forced to close down or operate at low capacity levels for lack of sufficient supplies of grain. The would-be miller must realise that he will be competing against other local mills and/or large-scale mills located in urban areas. Furthermore, in some countries, government legislation may favour the priority supply of maize grain to large-scale units, especially if these are publically-owned. Thus, depending on circumstances, the miller may need to adopt a lower scale of production than what is required by the estimated market demand as a result of an insufficient supply of maize.

Once the yearly output of maize meal has been determined, the miller will need to estimate the equipment capacity. The latter will be a function of the peak milling volume whenever the lack of grain storage capacity and/or market demand does not allow a constant milling capacity over the whole year. For example, a custom mill may need to process a large volume of maize in a relatively short time after the harvesting period and smaller volumes thereafter. A merchant mill, with a limited storage area, may also face similar conditions. On the other hand, a mill designed to process various types of grain harvested at different periods of the year, may maintain a constant daily output over the year.

The would-be miller must also decide whether he will carry out various ancillary operations (e.g. shelling, sifting, packaging). His decision will be mostly a function of market demand - which determines the quality, type and marketing of meal - and of the source of raw materials (e.g. shelled or unshelled maize). The capacity of the equipment used in these operations should correspond to that of the milling equipment.

III. METHODOLOGICAL FRAMEWORK FOR THE ESTIMATION OF MILLING COSTS

Once the yearly output of meal products and the capacity of various pieces of equipment have been determined, the would-be miller will need to estimate the milling cost per unit of output for each alternative milling technique suitable for the adopted scale of production. As indicated in the previous chapters, the miller may choose among various types of shelling and milling equipment and select those which minimise production costs and best suit local conditions. For example, the choice of a diesel-powered mill may be preferable to an electric one if electricity supply is not reliable even though production costs may be lower for the latter mill than for the former one.

The estimation of production costs per unit of output must take into consideration the following items:

- depreciation costs of equipment and buildings;
- maintenance and repair costs of equipment;
- energy costs;
- rental cost of land;
- labour costs;
- interest payments on working capital (in the case of merchant mills); and
- packaging costs (in the case of merchant mills).

The methodology for estimating the above cost items is described below.

III.1 Depreciation costs of equipment and buildings

The yearly depreciation costs of equipment and buildings are a function of the purchase prices of the above items, the prevailing interest rate and the useful life of the individual pieces of equipment and buildings. The higher the interest rate and the shorter the useful life, the higher will be the depreciation costs. These may be readily estimated as follows:

Let us assume that the purchase price of a piece of equipment (Z) is 10,000 dollars, that the prevailing interest rate is 15 per cent and that the useful life of the piece of equipment is 10 years. One may then estimate the yearly depreciation cost of this equipment with the help of table VI.5 which provides the present worth of an annuity factor for various interest rates (5 per cent to 40 per cent) and various periods of time (1 year to 25 years). The table shows that the factor (F) corresponding to an interest rate of 15 per cent and a life of 10 years is equal to 5.019. The yearly depreciation cost may then be easily calculated by dividing the purchase price Z (10,000 dollars) by the factor F (5.019 in this case).[1] In this example, the yearly depreciation cost is equal to :

$$\frac{Z}{F} = \frac{10,000}{5.019} = \underline{1,992 \text{ dollars}}$$

The value Z of the equipment or building may be estimated as follows:

---

[1] It may be shown that :

$$F = \frac{(1 + r)^n - 1}{r (1 + r)^n}$$

In the above equation, r is the prevailing interest rate and n is the useful life of the building or equipment in years. If one wishes to obtain a more accurate estimate of the yearly depreciation cost of equipment and buildings by taking into account their salvage value at the end of their useful lives, one may use the following equation:

$$\text{Yearly depreciation cost} = \frac{Z}{F} - \frac{S.r}{(1 + r)^n - 1}$$

where: Z, F, r and n are defined as above and S = salvage value of building or equipment.

<u>Value of buildings</u>: to obtain an estimate from local construction firms on the basis of the floor plans shown in Chapter V, and the adopted type and quality of buildings. Building costs will greatly vary according to the adopted quality standard. For example, low-cost buildings may be constructed at 50 dollars or less per m^2 while high standard buildings may cost as much as 400 dollars per m^2.

<u>Value of equipment</u>: The value of equipment must include transport and insurance costs whether they are imported or bought locally. Table VI.1 provides estimates of "on-site" costs of milling equipment with outputs ranging from 25 kg per hour (vertical stone mill) to 10,000 kg per hour (roller mill). Both the f.o.b. and c.i.f. costs are provided. Table VI.2 provides estimates of "on-site" costs of maize shellers with outputs ranging from 100 kg per hour (manual sheller) to 4,000 kg per hour (motor-driven shellers).

The above estimated costs of milling and shelling equipment may be lower or higher than the costs which will apply at the time the equipment is ordered, depending on the origin of the equipment, the type of transport used, the geographical location of the country, etc. The would-be miller is therefore advised to obtain accurate quotations from local importers of equipment or foreign suppliers (see selected list of equipment manufacturers in Appendix I). The cost of locally-produced equipment may be easily obtained from local manufacturers.

The would-be miller may also need to obtain estimates of the cost of various other pieces of equipment, such as sieving equipment, grain-cleaning equipment, packaging equipment, electric generators, etc. In some cases, some of the above equipment is attached to the mill itself.

The miller may obtain an estimate of the life of milling equipment from the manufacturer. In general, the life of milling equipment may exceed 50 years if it is properly operated, maintained and repaired.

III.2 Maintenance and repair costs of equipment

Maintenance and repair costs include the cost of labour and that of spare parts. If maintenance is carried out by the miller and/or a skilled mill worker, the labour costs for maintenance may be included int the overall labour costs of the mill (see section III.5). On the other hand, if the

Table VI.1

Estimated "on-site" costs of milling equipment

Mill type	Output kg/hr	Motor kW	Country of origin	F.O.B.	C.I.F. and Inland freight (end 1980)
1. Stone, vert.	25	0.5	Belgium	383	420
2. Stone, vert.	35	0.7	Belgium	401	440
3. Plate	45	0.7	United Kingdom	200	220
4. Stone, vert.	50	0.7	Belgium	466	510
5. Hammer	85	4.0	United Kingdom	2,040	2,250
6. Plate	125	2.0	United Kingdom	217	240
7. Stone, horiz.	150	3.0	France	770	850
8. Plate	150	3.5	India	900	1,000
9. Hammer	160	7.5	United Kingdom	1,826	2,000
10. Plate	180	3.5	United Kingdom	207	230
11. Hammer	180	7.5	United Kingdom	3,150	3,500
13. Stone, vert.	200	2.0	Denmark	277	300
14. Stone, vert.	240	2.0	Fed.Rep. Germany	624	690
15. Stone, horiz.	250	5.0	France	1,000	1,100
16. Plate	250	5.0	India	1,000	1,100
17. Plate	270	5.0	United Kingdom	260	290
18. Hammer	300	5.5	United Kingdom	1,124	1,250
19. Plate	300	3.0	France	205	225
20. Stone, vert.	310	3.0	Fed.Rep. Germany	654	720
21. Hammer	320	15.0	United Kingdom	3,387	3,725
22. Plate	340	5.5	United Kingdom	520	570
23. Stone, vert.	350	3.0	France	470	520
24. Stone, vert.	410	5.0	Fed.Rep. Germany	970	1,100
25. Hammer	500	5.5	Brazil	430	475
26. Hammer	570	7.5	United Kingdom	1,276	1,400
27. Stone, horiz.	600	7.5	Denmark	447	490
28. Hammer	680	30.0	United Kingdom	6,400	7,000
29. Stone, horiz.	700	9.0	Denmark	603	660
30. Stone, vert.	750	5.5	France	604	660
31. Stone, vert.	950	15.0	Denmark	683	750
32. Hammer	1,000	11.0	Brazil	750	825
33. Hammer	1,260	56.0	United Kingdom	10,400	11,450
34. Roller	2,000	110.0	United Kingdom	250,000	275,000
35. Roller	5,000	300.0	United Kingdom	400,000	440,000
36. Roller	10,000	485.0	United Kingdom	700,000	770,000

Source: Machinery makers and authors' estimates.

Table VI.2

Estimated "on-site" costs of maize shellers

Output kg/hr	Drive	kW	Country of origin	f.o.b. cost	c.i.f. and inland freight cost (mid 1981)
100	Manual	-	United Kingdom	26	30
150	Manual	-	Fed. Rep. Germany	35	40
350	Treadle	-	Japan	243	270
500	Manual	-	United Kingdom	125	140
600	Motor	1.5	Fed. Rep. Germany	906	1,000
750	Motor	0.3	United Kingdom	174	190
1,125	Motor	1.5	Japan	580	640
1,300	Pedal	-	United Kingdom	180	200
2,500	Motor	4.5	United Kingdom	453	500
3,000	Motor	7.5	United Kingdom	2,650	3,000
2,600	Motor	5.5	Brazil	1,250	1,400
4,000	Motor	7.5	United Kingdom	2,950	3,200

Source: Machinery manufacturers and authors' estimates.

---

the miller must secure regular maintenance services from local fitters, he will need to obtain an estimate of the yearly cost of these services.

The cost of spare parts is a function of the equipment yearly utilisation rate, the quality of the grain and the care with which the equipment is maintained. As a general rule, the yearly cost of spare parts may be estimated at 10 per cent of equipment cost.

III.3 Energy costs

The milling equipment may be powered by electric motors, diesel engines or petrol engines. The yearly cost of each of the above sources of energy may be estimated as follows:

- <u>Yearly cost of electricity</u>

This cost may be obtained from the following relationship:

Yearly cost of electricity =

$$(H) \times (A) \times (B) + C + D$$

where

H = estimated number of hours per year during which the equipment is functioning;

A = the rated kW of installed motor or motors;

B = the unit charge per kW

C = the yearly standing charge; and

D = other yearly charges.

It may be noted that the value of H is generally inferior to that of the total number of hours during which the mill is open since the equipment must be regularly stopped for maintenance and repairs.

- <u>Yearly cost of diesel or petrol</u>

This cost may be obtained from the following relationship:

Yearly cost of diesel or petrol = $(H) \times (F) \times (E)$

where : H = estimated number of hours per year

F = the hourly consumption of diesel or petrol (in litres)

E = the price per litre of diesel or petrol.

The above energy costs must be estimated for each engine-powered equipment (e.g. shellers, sieves, dryers). Tables VI.1 and VI.2 provide the engine power in kW for mills and shellers of various capacities. The calculation of the engine power in hp may be obtained by multiplying the rated power in kW by 1.3410.

III.4 <u>Rental cost of land</u>

Whether the land is owned by the miller or not, a yearly rental cost of land must be estimated and added to the other cost items. The yearly rental cost may be estimated on the basis of the required floor area (see Chapter V) and the prevailing rental rate for land at similar locations.

III.5 <u>Labour costs</u>

The yearly labour costs are a function of the mill's capacity, the type of mill (e.g. custom, merchant or roller mill) and the number of sub-processes which are carried out (e.g. storage, shelling, sifting, packaging, maintenance). The operation of a small-scale custom mill may be carried out by the mill owner and, in some cases, a helper since the milling of grain is the only operation. On the other hand, a small-scale or intermediate-scale merchant mill will require a full-time or part-time manager (the mill owner in most cases), a part-time or full-time skilled worker for the operation and maintenance of the equipment (the mill owner may carry out these tasks), and a number of unskilled workers for the moving of grain and product, packaging, cleaning, etc.

The number of unskilled workers needed to move the grain and the product depends on whether the output is stored by the mill prior to despatching to clients. If such storage does take place, labour is needed for moving the grain from the trucks to the storage area and from the latter to the mill as needed. The product (i.e. the meal) must then be moved from the mill to the product storage area, and finally from the latter to the trucks. Thus, each tonne of maize (in the form of grain or product) must be moved four times. The number of unskilled workers (w) needed to perform the above tasks is then equal to:

$$w = \frac{4 \times m}{p}$$

where : m = number of tonnes of maize processed per day; and
        p = productivity of labour, in tonnes per day.

The value of p is a function of the local labour productivity, of the distance between the trucks and the storage area, and of the idle time between the unloading of trucks. As the above factors will vary significantly from one mill to another, no attempt will be made to provide an estimate of p. The latter must be obtained by the mill's owner on the basis of information at his disposal.

Labour may also be required for the shelling of cobs and the bagging or packing of the product. The number of unskilled workers needed for the above operations is a function of the mill's capacity. In the case of small-scale or medium-scale merchant mills, one or two operators should be able to carry out each of the above operations.

The number of unskilled workers may vary over the year, depending on whether the mill stores sufficient grain for the whole year or not. In any case, an estimation of unskilled labour requirements should take into consideration the possible closing down of the mill over a certain period of the year.

The total yearly labour and managerial costs (L) may be calculated from the following relationship:

$$L = w_u.d.s_u + 12.w_s.S + 12.M$$

where:

w_u = number of unskilled workers per day for the moving of grain and product, bagging or packaging, and cleaning;

d = number of days in the year during which unskilled labour is employed;

s_u = daily wage of unskilled labour;

w_s = number of skilled workers (assumed to be part of the permanent staff);

S = monthly salary of skilled labour;

M = monthly salary of mill owner.

The above formulation may be adapted to take into consideration the following variations:

- no moving costs of grain and product in the case of custom mills;
- no packaging costs; and
- no skilled labour costs if the work is done by the mill's owner.

The yearly salary of the mill's manager should be at least equivalent to that he would get in his next best alternative employment. The salaries and wages of skilled and unskilled labour are those prevailing in the mill's location.

III.6 Packaging costs

Maize meal may be bagged in 50 kg jute sacks or smaller sacks (e.g. 5 kg or more) depending on the local marketing conditions. The meal may also be marketed in 1 kg paper bags bearing the appropriate information (e.g. brand name, net weight, name of the mill). The jute sacks or paper bags will generally be ordered from local manufacturers. The yearly packaging costs are equal to the value of the yearly order of sacks and/or paper bags required for the packaging of the total yearly output of the mill.

III.7 Interest payments on working capital

Merchant mills may often need to store a volume of grain equivalent to three months' operation of the mill. Such storage may not be avoided given the uneven supply of maize grain over the year. Similarly, they may also need to store the produced meal and/or agree to delayed payments from clients. Thus, a volume of meal equivalent to a one month's or two months' operation of the mill may be either kept as a stock and/or sold on credit. Consequently, the miller needs some working capital to cover the cost of a volume of grain equivalent to four to five months' operations (grain in stock and grain in the form of meal) and that of labour, energy, and equipment depreciation for the processing of the stocked meal. Since this working capital may be considered idle capital, the yearly interest payments on the latter constitute a cost item to be added to the other items.

III.8 Unit production cost of meal
produced by small-scale mills

The estimation of unit production cost of maize meal is the last step in the evaluation of alternative maize milling techniques. The unit production cost is equal to the sum of the yearly cost items (identified in sections III.1 to III.7) divided by the total yearly output of maize meal. The most appropriate milling technique is that associated with the lowest unit production cost. The choice of equipment and scale of production should therefore correspond to the identified lowest cost technique.

In the case of custom mills, the most appropriate milling technique will be the one associated with the lowest sum of depreciation costs (building and equipment) and energy cost per unit of output since labour inputs are the same for all techniques (i.e. the labour of the mill's owner and, in some cases, that of a helper).

In the case of merchant mills, the evaluation of alternative milling techniques should take into account market demand for various qualities of meal. For example, a miller may market either non-processed whole meal or partly sifted whole meal (i.e. partial removal of germ and bran). In the latter case, he may use a mill equipped with a sifting device. The comparison between the production of whole meal and that of partly sifted meal should therefore take into consideration the differential cost of the equipment and the differential retail prices of the product (i.e. the price of whole meal and that of partly sifted meal) and of by-products (i.e. bran and germ). In this case, the choice of technology involves both a choice of milling technique and a choice of product.

III.9 Unit production cost of meal produced by roller mills

The cost structure of roller mills is much more complex than that of small-scale mills. Since this memorandum is mostly concerned with these latter mills, no attempt will be made to describe in detail the cost structure of roller mills.

The principal cost items for a roller mill are those of raw maize, packing, labour, storage of raw materials and products, management and administration, transport, insurance and taxes. Other important cost determinants are location, working capital tied up in raw maize storage, capacity utilisation and length of operating season. According to Uhlig and Bhat (1980), raw maize accounts for "more than 80 per cent of total net present costs discounted at 10 per cent" on the basis of Kenyan factor prices. This figure provides a very convenient, if rough guide to the estimation of unit costs. One only needs to know the local price of raw maize in order to estimate unit production costs. If a more accurate guide is needed (e.g. in cases where estimated unit costs are close to retail prices), more detailed information is needed on the quantity and price of the factors of production.

The cost of packing materials ranks second in magnitude to raw maize costs. It represents 6 to 9 per cent of the total unit cost depending on whether the packing of meal is done manually or mechanically. Packing costs can be lowered by a small percentage if larger bags or sacks are used (e.g. use of 50 kg sacks instead of 25 kg sacks).

Labour and management costs rank third at under 5 per cent of unit costs, reaching 2 per cent in machine-intensive plants. The management proportion of this cost element is high, approximately 48 per cent of total labour and management costs for the smalllest scale roller mills (2 tonnes/hour). This proprotion decreases as the scale of production increases, but does not generally go below 32 per cent.

Other cost items for roller mills are far less significant than the three cost items indicated above. Insurance, administration and maintenance costs represent approximately 4 per cent of total unit costs.

IV. ILLUSTRATIVE EXAMPLES OF THE ESTIMATION PROCEDURE

The estimation procedure described in section III may be illustrated with respect to two types of mills: a custom mill with an output of 1 tonne per day and a small-scale merchant mill with an output of 8 tonnes per day. The custom mill uses an electric plate mill equipped with a 2 kW motor (mill No. 6 in table VI.1), while the merchant mill uses an electric hammer mill equipped with an 11 kW motor (mill No. 32 in table VI.1). Both mills produce whole meal.

Table VI.3 provides the various yearly cost items estimated on the basis of the following assumptions:

(i) Custom mill (rural area)
 - Output : 1 tonne per day;
 - Organisation of production : 8 hours per day, 300 days per year;
 - Price of maize : US$150 per tonne of grain;
 - Price of plate mill (on-site) : US$500;
 - Cost of building : US$300;
 - Labour : Mill's owner and assistant;
 - Energy cost : US$0.1 per kWh;
 - Energy consumption : 2,400 kWh per year;
 - Spare parts : US$50 per year (10 per cent of equipment cost);
 - Land rental : US$300 per year;
 - Life of building and equipment : 25 years;
 - Interest rate : 15 per cent;
 - Working capital : none;
 - Mill's yearly profits : US$5,000; and
 - Monthly salary of assistant : US$150.

(ii) <u>Merchant mill</u> (urban area)

- Output : 8 tonnes per day;
- Organisation of production : 8 hours per day; 300 days per year;
- Price of maize : US$150 per tonne of grain;
- Price of hammer mill : US$1,200;
- Cost of buildings : US$4,000;
- Labour : Mill's owner plus four unskilled workers;
- Energy costs : US$0.1 per kWh;
- Energy consumption : 26,400 kWh per year;
- Spare parts : US$120 per year;
- Land rental : US$1,200 per year;
- Life of buildings and equipment : 25 years;
- Interest rate : 15 per cent;
- Working capital : Volume of maize grain equivalent to one or three
months of mill's operation (200 tonnes or 600 tonnes);
- Mill's yearly profits : US$12,000; and
- Monthly salary of unskilled labour : US$200.

It is assumed that the maintenance of both mills is carried out by the mill's owner. Bagging or packaging costs are not considered in this example in order to facilitate the comparison between the two types of mills. However, it is most probable that the merchant mill will need to add a packaging cost to the other cost items.

Table VI.4 provides estimates of unit production costs for both mills as well as a minimum wholesale price of maize meal. It may be seen that, in this particular example, the unit production cost for the merchant mill is significantly lower than that for the custom mill even though the merchant mill must keep a stock of grain in order to avoid temporary shutdowns of the mill. In case the merchant mill does not need to maintain a stock of grain, the unit production cost will be US$11 per tonne instead of US$13 or US$17. It may also be noted that unit production costs vary between 7.98 per cent and 14.30 per cent of the wholesale price. These fractions are significantly lower than the approximately 20 per cent which applies to large-scale roller mills. However, one should not compare roller mills with small-scale custom or merchant mills since the type of output produced by the latter is different from that produced by roller mills.

Table VI.3

Yearly costs of a custom mill and of a merchant mill

	in US$	
	Custom mill	Merchant mill
Labour costs	1,800	9,600
Profits	5,000	12,000
Equipment and building depreciation costs[1]	124	805
Energy costs	240	2,640
Spare parts	50	120
Land rental costs	300	1,200
Interest on working capital[2]	-	4,500; 13,500
Total yearly costs[2]	7,514	30,865; 39,865

Table VI.4

Unit production costs of a custom mill and of a merchant mill

	Custom mill	Merchant mill
Total yearly costs[2] (US$)	7,514	30,865; 39,865
Total yearly output (tonnes)	300	2,400
Unit production cost[2] (US$ per tonne)	25	13 ; 17
Minimum wholesale price of meal[2] (US$ per tonne)	175	163 ; 167
Production cost as a percentage of wholesale price (per cent)	14.30	7.98; 10.18

[1] Factor F is equal to 6.464 (see table VI.5);

[2] Estimated on the basis of one month's stock and three months' stocks of grain (merchant mill only).

This illustrative example should not lead to the conclusion that unit production costs are lower for merchant mills than for custom mills. The example is based on too many assumptions which may not apply in a large number of cases. It is therefore essential that the would-be miller undertakes his own evaluation, based on accurate estimates of the various cost items, with a view to identifying the milling technique which is the most suitable to prevailing local conditions.

Table VI.5

Year	5%	6%	8%	10%	12%	14%	15%	16%	18%	20%	22%	24%	25%	26%	28%	30%	35%	40%
1	0.952	0.943	0.926	0.909	0.893	0.877	0.870	0.862	0.847	0.833	0.820	0.806	0.800	0.794	0.781	0.769	0.741	0.714
2	1.859	1.833	1.783	1.736	1.690	1.647	1.626	1.605	1.566	1.528	1.492	1.457	1.440	1.424	1.392	1.361	1.289	1.224
3	2.723	2.673	2.577	2.487	2.402	2.322	2.283	2.246	2.174	2.106	2.042	1.981	1.952	1.923	1.868	1.816	1.696	1.589
4	3.546	3.465	3.312	3.170	3.037	2.914	2.855	2.798	2.690	2.589	2.494	2.404	2.362	2.320	2.241	2.166	1.997	1.849
5	4.330	4.212	3.993	3.791	3.605	3.433	3.352	3.274	3.127	2.991	2.864	2.745	2.689	2.635	2.532	2.436	2.220	2.035
6	5.076	4.917	4.623	4.355	4.111	3.889	3.784	3.685	3.498	3.326	3.167	3.020	2.951	2.885	2.759	2.643	2.385	2.168
7	5.786	5.582	5.206	4.868	4.564	4.288	4.160	4.039	3.812	3.605	3.416	3.242	3.161	3.083	2.937	2.802	2.508	2.263
8	6.463	6.210	5.747	5.335	4.968	4.639	4.487	4.344	4.078	3.837	3.619	3.421	3.329	3.241	3.076	2.925	2.598	2.331
9	7.108	6.802	6.247	5.759	5.328	4.946	4.772	4.607	4.303	4.031	3.786	3.566	3.463	3.366	3.184	3.019	2.665	2.379
10	7.722	7.360	6.710	6.145	5.650	5.216	5.019	4.833	4.494	4.192	3.923	3.682	3.571	3.465	3.269	3.092	2.715	2.414
11	8.306	7.887	7.139	6.495	5.938	5.453	5.234	5.029	4.656	4.327	4.035	3.776	3.656	3.544	3.335	3.147	2.752	2.438
12	8.863	8.384	7.536	6.814	6.194	5.660	5.421	5.197	4.793	4.439	4.127	3.851	3.725	3.606	3.387	3.190	2.779	2.456
13	9.394	8.853	7.904	7.103	6.424	5.842	5.583	5.342	4.910	4.533	4.203	3.912	3.780	3.656	3.427	3.223	2.799	2.468
14	9.899	9.295	8.244	7.367	6.628	6.002	5.724	5.468	5.008	4.611	4.265	3.962	3.824	3.695	3.459	3.249	2.814	2.477
15	10.380	9.712	8.559	7.606	6.811	6.142	5.847	5.575	5.092	4.675	4.315	4.001	3.859	3.726	3.483	3.268	2.825	2.484
16	10.838	10.106	8.851	7.824	6.974	6.265	5.954	5.669	5.162	4.730	4.357	4.033	3.887	3.751	3.503	3.283	2.834	2.489
17	11.274	10.477	9.122	8.022	7.120	6.373	6.047	5.749	5.222	4.775	4.391	4.059	3.910	3.771	3.518	3.295	2.840	2.492
18	11.690	10.828	9.372	8.201	7.250	6.467	6.128	5.818	5.273	4.812	4.419	4.080	3.928	3.786	3.529	3.304	2.844	2.494
19	12.085	11.158	9.604	8.365	7.366	6.550	6.198	5.877	5.316	4.844	4.442	4.097	3.942	3.799	3.539	3.311	2.848	2.496
20	12.462	11.470	9.818	8.514	7.469	6.623	6.259	5.929	5.353	4.870	4.460	4.110	3.954	3.808	3.546	3.316	2.850	2.497
21	12.821	11.764	10.017	8.649	7.562	6.687	6.312	5.973	5.384	4.891	4.476	4.121	3.963	3.816	3.551	3.320	2.852	2.498
22	13.163	12.042	10.201	8.772	7.645	6.743	6.359	6.011	5.410	4.909	4.488	4.130	3.970	3.822	3.556	3.323	2.853	2.498
23	13.489	12.303	10.371	8.883	7.718	6.792	6.399	6.044	5.432	4.925	4.499	4.137	3.976	3.827	3.559	3.325	2.854	2.499
24	13.799	12.550	10.529	8.985	7.784	6.835	6.434	6.073	5.451	4.937	4.507	4.143	3.981	3.831	3.562	3.327	2.855	2.499
25	14.094	12.783	10.675	9.077	7.843	6.873	6.464	6.097	5.467	4.948	4.514	4.147	3.985	3.834	3.564	3.329	2.856	2.499

Present worth of an annuity factor

How much 1 received or paid annually for X years is worth today

APPENDICES

APPENDIX I

EQUIPMENT MANUFACTURERS AND SUPPLIERS

The equipment manufactured and/or supplied by the firms listed below is designated by the following letters:

- HH   : Husking hooks
- RHS : Rotary hand shellers
- FSHS: Free standing hand shellers
- PS   : Powered shellers
- PM   : Plate mills
- HM   : Hammer mills
- SM   : Stone mills
- RM   : Roller mills
- VFD : Ventilated floor dryers
- ISD : In-storage dryers
- BD   : Batch dryers
- CD   : Continuous dryers

Manufacturers/Suppliers	Equipment
BELGIUM	
DDD President, Chaussée de Dikkebus, 487, 8904 YPRES	SM
BRAZIL	
Irmaos Nogueira SA., CIMAG Ltda, Av. Ipiranga 1071, SAO PAULO	PS
Laredo S.A., Rua 1 de Agosto, 11-67 CEP 17100 BAURU (SP)	PS, HM
DENMARK	
ABC Hansen Comp. A/S., P.O. Box 3054, DK 1508 COPENHAGEN V	PM, SM
Erling Foss Export, Thorsgade 59, DK-2200 COPENHAGEN N	SM

United Milling Systems Ltd.,
8 Gamle Carlsberg vej,   RM
DK-2500 VALBY;COPENHAGEN

Skjold,
P.O. Box 39,   HM, SM
DK 9300 SAEBY

FEDERAL REPUBLIC OF GERMANY

AMOS Machinenfabrik Gmbh,
Postfach 1160   RHS, PS
D-7100 HEILBRONN

Iruswerke Dusslingen
J. Rilling & Sohne,   SM
Postfach 128,
D-7401 DUSSLINGEN

FRANCE

Argoud SECA
Le Mottier   PM, HM
38260 LA COTE ST-ANDRE

Ets. Champenois, S.A.,
Chamouilley,   RHS, FSHS, SM
52170 CHEVILLON

Ets. Claudien Beroujon,
280, rue des Alpes,   SM
38290 LA VERPILLIERE

Société COMIA-FAO, S.A.,
27, bd de Chateaubriand,   PS, HM, SM
35500 VITRE

Electra,
Poudenas,   HM
47170 MEZIN

Ets. A. Gaubert,
22, rue Gambetta,
B.P. 24,   SM
16700 RUFFEC

Goudard,
77260 LA FERTE SOUS JOUARD   HM

Law SECEMIA,
B.P. 15,
5, ave. du Général de Gaulle,   HM, RM
60304 SENLIS CEDEX

Moulis,
80800 MONTREDON LA BESSONIE   PM

PROMILL,
B.P. 109
28104 DREUX
HM

Renson et Cie.,
B.P. 23,
59550 LANDRECIES
RHS, FSHS, PS, PM, SM

SAMAP,
1, rue du Moulin,
B.P. 1 Andolsheim
68600 Neufbrisach
PM

Ets. Simon Frères,
Rue Laurent Simon
B.P. 171,
50104 CHERBOURG CEDEX
SM

Tixier Frères,
18120 LURY SUR ARNON
PM, HM

HONG KONG

China National Machinery Import
 and Export Corporation,
c/o China Resources Company,
Bank of China Building,
HONG KONG
RHS

INDIA

Allied Trading Company,
Railway Road,
AMBALA CITY 134 002
RHS, PS

Binny Engineering Company,
Meenambakkam,
P.O. Box 1111,
MADRAS 600 001
RM

Cossul & Co. Ltd.,
Industrial Area 123/367,
Fazalgang,
KANPUR
RHS, FSHS

Dandekar Brothers,
Shivagi Nagar
SANGLI (Maharashtra)
RHS

Dandekar Machine Works,
Bhiwandi,
1-421 302 Dist. Thana,
MAHARASHTRA
RHS, PS, SM

International Manufacturing Co.,
Hospital Road,
Jagraon,
LUDHIANA (Punjab)
PS

Kisan Krishi Yantra Udyog,
64 Moti Bhawan   HM, PM
Collectorganj,
KANPUR 208 001

Numex Engineers,
P.O. Box 820,   PM
BOMBAY 400 001

Rajan Trading Co.,
P.O. Box 250,   PM
MADRAS 600 001

Rajasthan State Agro Industries Corp., Ltd.,
Virat Bhawan, C-Scheme,   RHS
JAIPUR 302 006 (Rajasthan)

Union Forgings,
GT Road,
Focal Point,   FSHS
Shepur
LUDHIANA

ITALY

Ceccato Olindo Machine Snc.,
Via Giustiniani 1,   RHS, FSHS, PS
35010 ARSEGO (PD)

Favini & Co.,
Industria Meccanica,   RM
Via Provinciale 13,
Fornova,
BERGAMO

OCRIM Spa.,
Via Massarotti No. 76,   RM
CREMONA

IVORY COAST

S.A.C.M.
16, rue des Foreurs,
B.P. 4019   PM, HM
ABIDJAN

JAPAN

CeCoCo Ltd.,
P.O. Box 8,   RHS, FSHS
IBARAKI CITY, OSAKA 567

KENYA

Ndume Products Ltd.,
P.O. Box 62,   HM
GILGIL

MALAWI

Brown and Clapperton Ltd.,
P.O. Box 52   RHS, HM
BLANTYRE

NETHERLANDS

Ten Have Engineering bv
Industrieweg 11,   HM
Postbus 27,
7250AA VORDEN

PHILIPPINES

Guanko Ironworks,
102-104 Mj Cuenco Avenue,   RM
CEBU CITY

SENEGAL

SISMAR (Ex. SISCOMA)
Rue du Dr. Theze & Grammont   FSHS, HM
B.P. 3214
DAKAR

D. Seck,
KEBEMER   HM

C. Gueye
GOSSAS   HM

SWITZERLAND

Buhler Bros Ltd.,   RM
CH-9240 UZWIL 4

TANZANIA

Ubungo Farm Implements,
P.O. Box 2669   RHS
DAR ES SALAAM

UNITED KINGDOM

Alvan Blanch Development Co., Ltd.,
Chelworth   PS, PM, HM, VFD
MALMESBURY, Wiltshire SN16 9SG

E.H. Bentall & Co. Ltd.,
MALDON, Essex CM9 7NW   PM, HM, VFD, ISD, CD

Christy & Norris Ltd.,
Broomfield Road,   HM
CHELMSFORD CM1 1SA

Colman & Co. (Agricultural) Ltd.,
Ballingdon Works,   VFD, BD
SUDBURY, Suffolk CO10 6BY

Cornercroft (Agriculture) Ltd.,
CONINGSBY, Lincs LN4 4SN   VFD

R. Hunt and Co. Ltd.,
Atlas Works,   RHS, FSHS, PM
Earls Colne,
COLCHESTER, Essex CO6 2EP

Kamas Machinery Ltd.,
110 Hunslet Lane,   CD
LEEDS, Yorkshire LS10 1ES

Law-Denis Engineering Ltd.,
Lavenham Road,
The Beeches Industrial Estate,   ISD, CD
Yate,
BRISTOL BS17 5QX

R.A. Lister Ltd.,
Dursley   VFD, ISD
GLOS GL11 4HS

Miracle Mills, Ltd.,
Franklin Road,   HM
LONDON SE20 8JD

Ransomes, Sims & Jeffries Ltd.,
IPSWICH, Suffolk IP3 9QG   PS, FSHS

Scotmec Ltd.,
42-44 Waggon Road,   HM
AYR (Scotland)

Henry Simon Ltd.,
P.O. Box 31,   HM, RM
STOCKPORT, Cheshire SK3 ORT

Turner Grain Handling Ltd.,
Benezet Street,   CD
IPSWICH, Suffolk 1P1 2JQ

John Wilder (Engineering) Ltd.,
Hithercroft Works,   BD
WALLINGFORD, Oxon OX10 9AR

UNITED STATES

Bryant - POFF Incorp.,
P.O. Box 127,   PS
COATESVILLE, Indiana 46121

Jacobson International Inc.,
2445 Nevada Avenue North,   HM
MINNEAPOLIS, Minnesota 55427

Raidt Manufacturing Co.,
SHENANDOAH, Ohio   HH

Seedburo Equipment Co.,
1022 West Jackson Blvd.,   PS, RHS
CHICAGO, Illinois 60607

Seedburo Equipment Co.,
1022 West Jackson Blvd.,   PS, RHS
CHICAGO, Illinois 60607

APPENDIX II

SELECTED RESEARCH, ACADEMIC AND APPROPRIATE TECHNOLOGY
INSTITUTIONS INVOLVED IN GRAIN PROCESSING

BRAZIL

Fundaóao Centro Tecnologico de Minas Gerais,
MINAS GERAIS

CANADA

Canadian Hunger Foundation,
75, Sparks St.,
OTTAWA, Ontario KIP 5A5

ECUADOR

Centro de Desarrollo Industrial
de Ecuador,
GUAYAQUIL

ETHIOPIA

International Livestock Centre for Africa,
ADDIS ABABA

FRANCE

Groupe de Recherches et Echanges Technologiques,
34, rue Dumont d'Urville,
75116 PARIS

INDIA

Central Food Technological Research Institute,
MYSORE 570 013

International Centre for Research in the
Semi-Arid Tropics
HYDERABAD

Protein Foods & Nutrition Development
Association of India,
BOMBAY

Agricultural Research Institute,
NEW DELHI

JAMAICA

Caribbean Food and Nutrition Institute
KINGSTON

MEXICO

International Maize and Wheat Improvement
Centre,
LONDRES

NETHERLANDS

Technische Hogeschool,
EINDHOVEN

NIGERIA

International Institute of Tropical Agriculture,
IBADAN

PAKISTAN

IRRI-PAK Agricultural Machinery Programme,
ISLAMABAD

Pakistan Agricultural Research Council,
ISLAMABAD

PHILIPPINES

National Grain Authority
QUOXON CITY

THAILAND

Asian Institute of Technology,
P.O. Box 2754,
BANGKOK

Thailand Institute for Scientific &
Technological Research
BANGKOK

TRINIDAD AND TOBAGO

Ministry of Agriculture, Lands & Fisheries,
PORT-OF-SPAIN

UNITED STATES

Volunteers in Technical Assistance,
1815 North Lynn Street,
Suite 200,
ARLINGTON, Virginia 22209

ZAMBIA

Northern Technical College,
NDOLA

ZAIRE

CEDECO,
B.P. 70;
KIMPESE

APPENDIX III

BIBLIOGRAPHY

Ajayi, O.A. : Grinding in hammer mills, Report of the National College of Agricultural Engineering (Silsoe, United Kingdom, NCAE, 1980).

Beaty, H.H.; Shore, G.C. : Drying shelled corn, Circular No. 916, Cooperative Extension Service (Urbana, Illinois, University of Illinois, College of Agriculture, 1965).

Berger, J.: Maize production and the maturing of maize (Geneva, Centre d'Etude de l'Azote, 1962).

Brekke, O.L. : "Corn dry milling: Pre-tempering low moisture corn", in Cereal Chem. (1967), Vol. 15, No. 44, pp. 521-531.

_____ : "Corn dry milling: Cold tempering and de-germination of corn of various initial moisture contents", in Cereal Chem. (1969), Vol.5, No. 46, pp. 545-549.

_____ : "Dry milling artificially dried corn: Roller milling of de-germinator stock at various moistures", in Cereal Science Today (1970), Vol. 2, No. 15, pp. 37-42.

_____ : "Dry milling of opaque - 2 (high lysine) corn", in Cereal Chem. (1971), Vol. 5, No. 48, pp. 499-511.

Bressani, R.; Castillo, S.V.; Guzman, M.A. : "Corn flours: The nutritional evaluation of processed whole corn flours", in Journal of Agricultural Food Chemistry (1962), Vol. 4, No. 10, pp. 308-312.

Brooker, D.B.; Bakker-Arkema, F.W.; Hall, C.W. : Drying of cereal grains (Westport, Connecticut, Avi Publishing, 1974).

Buelow, F.: Drying crops with solar heated air, Proceedings of a United Nations Conference on New Sources of Energy held in Rome in 1961 (Rome, FAO, 1961).

Christensen, C.M.; Kaufmann, H.H.: Grain storage: the role of fungi in quality loss (Minneapolis, Minn., University of Minnesota Press, 1969).

Chambers : A dictionary of science and technology (Edinburgh, Scotland, W & R Chambers Ltd., 1974).

Christian Council of Tanzania : Manually-operated grinding mills: An evaluation (Dar-es-Salaam, Department for Development Services, 1980).

Clarke, B. : A survey of cereal grinding in Africa. Report No. 190326 (Silsoe, United Kingdom, National College of Agricultural Engineering, 1980).

Commonwealth Secretariat: Guide to technology transfer in East, Central and Southern Africa (London, Commonwealth Secretariat, 1981).

Food and Agriculture Organisation of the United Nations : Maize and maize diets: A nutritional survey. Nutritional Studies No. 9 (Rome, 1953).

_____: Food composition tables for use in Africa (Rome, 1968).

_____: "Food composition tables - Minerals and vitamins for international use", in Nutritional Studies, 1954, No. 11.

_____: Production yearbook (Rome, 1979).

_____: Processing and storage of foodgrains by rural families (Rome, 1979).

Gerstenkorn, P.; Swingelberg : "Maize conditioning and milling" in Mühle und Mischfuttertechnik, 1975, Vol. 6, No. 112, pp. 66-70.

Groupe de recherches et d'échanges technologiques: Fichier technique du díveloppement, fascicule No.29 (Paris, 1983).

Harper, M.: Working Paper 174 (Nairobi, Institute for Development Studies, Nairobi University, 1974).

Inglett, G.E, : Corn: Culture, processing, products (Westport, Connecticut, Avi Publishers, 1970).

Intermediate Technology Development Group: Tools for agriculture - A Buyers' guide to low-cost agricultural implements, compiled by John Boyd (London, Intermediate Technology Publications, 1976).

Jobs and Skills Programme for Africa : Technologies appropriées dans les industries de transformation alimentaire et de conservation de fruits dans quatre pays de la CEAO - Haute-Volta, Mali, Niger, Sénégal (Addis-Ababa, ILO, 1982).

JASPA : Appropriate technologies in cereal milling and fruit processing industries - A comparative sub-regional study of four East African countries - Kenya, Somalia, Tanzania and Zambia (Addis-Ababa, ILO, 1981).

James, A.W. : Personal Communication (1982).

Kent, N.L.: Technology of cereals (London, Pergamon Press, 1966)

Lindblad, C.; Druben, L.: Small farm grain storage (Arlington, Virginia, VITA, 1977), vol. I, II and III.

Kaplinsky, R.: "A country case study : Food processing in Kenya", in C.G. Baron (ed.): Technology, Employment and Basic Needs in Food Processing in Developing Countries (Oxford, Pergamon Press, 1980).

Kent-Jones, D.W.; Amos, A.J. : Modern cereal chemistry (Sixth Ed.) (London, Food Trade Press, 1967).

Ndambuki, W.W.: "Meet Kenya's very own energy pioneer", in Daily Nation (Nairobi), 30 Oct. 1981.

O'Kelly, E.: "Processing and storage of foodgrains by rural families", in Agricultural Services Bulletin (Rome, FAO, 1979).

Olatunji, O.; Edwards, C.; Koleoso, O.A. : "Processing of maize and sorghum in Nigeria for human consumption" in Journal of Food Technology, 1980, Vol. 1, No. 15, pp. 85-92.

Schlage, C.: "Polished versus whole maize : Some nutritional and economic implications of the traditional processing of maize in North Eastern Nigeria", Research Report No. 2 (Dar-es-Salaam, Bureau of Resource Assessment and Land Use Planning, 1968).

Soza, R.F.; Willena, D. : Solar grain dryers. Paper presented at the 25th annual meeting of the Central American Cooperative Programme for the Improvement of Food Cultivation held in Tegucicalpa in 1979 (Honduras, PCCMCA, 1979).

Schoonhoven, A.V.; Horber, E.; Mills, R.B. : "Conditions modifying expression of resistance in maize kernels to the maize weevil" in Environmental Entomology, 1976, No. 5.

Stewart, F. : "The choice of technique: Maize grinding in Kenya", in Technology and underdevelopment (London, Macmillan, 1977).

Temple, M. : "Nepalese water mill", in Appropriate Technology (London), 1974, Vol. 1, No. 3, p. 15.

Tropical Products Institute: A wooden hand-held maize sheller (London, TPI, 1977).

Uhlig, S.J.; Bhat, B.A. : The choice of technique in maize milling (Edinburgh, Scotland, Scottisch Academic Press, 1979).

Vojnovich, C.; Pfeifer, V.F.; Griffin, E.L.: "Reducing microbial populations in dry milled corn products", in Cereal Science Today, 1978, Vol. 12, No. 15, pp. 401-407.

Waelti, H.; Buckle, W.F. : " Factors affecting corn kernel damage in combine cylinders", in Transactions of the American Society of Agricultural Engineers, 1967.

Walker, D.J. : Report of the Swaziland rural grain storage project - 1972 - 1975 (Mbabane, Ministry of Agriculture, 1975).

Wells, G.H.: "The dry side of corn milling", in Cereal Foods World, 1979, Vol. 8, No. 24, pp. 333-340.

Zambia Industrial and Minining Corporation Ltd. : Maize meal, maize milling: A national problem (Lusaka, 1978).

GLOSSARY OF TECHNICAL TERMS

Alkali	A soluble hydroxide of a metal
Ambient	Temperature of the surrounding air
Amylopectin	A branched-chained polymer of glucose
Amylose	A straight-chained polymer of glucose
Aspiration	The removal of small particles by suction
Carborundum	Trade name designating a proprietary range of products, e.g. silicon carbide
Corneous	Resembling horn in texture
Cyclone	Conical vessed used to extract dust by centrifugal action
Decortication	Removal of the outer layers from the grain
Emery	A finely granulated admixture of corundum (oxide of aluminium) and either magnetite or haematite
Germination	The beginning of growth of the embryo of a seed
Inflorescence	In flowering plants, the part of the shoot which bears the flowers
Monoecious	Having separate male and female flowers on the same individual plant

Pellagra	A chronic disease due to general dietary deficiency. It is characterised by gastro-intestinal disturbances, a redness of the skin, mental depression and paralysis
Pericarp	Wall of a fruit or seed, if derived from the ovary wall
Plumule	The rudimentary shoot in the seed embryo
Radicle	The rudimentary root in the seed embryo
Tempering	The addition of water to grain to increase its moisture content
Testa	The seed coat, several layers of cells in thickness derived from the outer layers of the ovule
Tiller	A branch produced from the base of a stem
Viscosity	The resistance of a fluid to flow
Whole meal	The milled product containing all parts of the grain

QUESTIONNAIRE

1. Full name...

2. Address...
   ....
   ....

3. Profession (check the appropriate case)

   Established maize miller../__/
   If yes, indicate scale of production................................

   Government official../__/
   If yes, specify position..

   Employee of a financial institution............................/__/
   If yes, specify position..

   University staff member../__/

   Staff member of a technology institution......................../__/
   If yes, indicate name of institution..............................
   ....

   Staff member of a training institution........................./__/
   If yes, specify...
   ....

   Other, specify...
   ....

4. From where did you get a copy of this memorandum?
   Specify if obtained free or bought...................................
   ....

5.   Did the memorandum help you achieve the following:
     (Check the appropriate case)

     Learn about maize milling techniques you were not aware of   /__/

     Obtain names of equipment suppliers   /__/

     Estimate unit production costs for various scales
     of production/technologies   /__/

     Order equipment for local manufacture   /__/

     Improve your current production technique   /__/

     Cut down operating costs   /__/

     Improve the quality of produced maize meal   /__/

     Decide which scale of production/technology to
     adopt for a new milling plant   /__/

     If a Government employee, to formulate new measures
     and policies for the maize milling industry   /__/

     If an employee of a financial institution, to assess
     a request of a loan for the establishment of a maize
     milling plant   /__/

     If a trainer in a training institution, to use the
     memorandum as a supplementary training material   /__/

     If an international expert, to better advise counter-
     parts on milling technologies   /__/

6.   Is the memorandum detailed enough in terms of: Yes   No

     - Description of technical aspects.........................____.....____

     - Names of equipment suppliers............................____.....____

- Costing information..____.....____

- Information on socio-economic impact....................____.....____

- Bibliographical information.............................____.....____

If some of the answers are 'No', please indicate why below or on a separate sheet:

....
....
....

7. How may this memorandum be improved if a second edition is to be published?...
....
....

8. Please send this questionnaire, duly completed to:

> Technology and Employment Branch
> International Labour Office
> CH-1211 GENEVA 22 (Switzerland)

9. In case you need additional information on some of the issues covered by this memorandum, the ILO and UNIDO would do their best to provide the requested information.